SpringerBriefs in Electrical and Computer Engineering

T0234322

For further volumes:
http://www.springer.com/series/10059

Kyung-Hyan Yoo · Ulrike Gretzel
Markus Zanker

Persuasive Recommender Systems

Conceptual Background and Implications

 Springer

Kyung-Hyan Yoo
William Paterson University
Wayne, NJ
USA

Markus Zanker
Alpen-Adria-Universität Klagenfurt
Klagenfurt
Austria

Ulrike Gretzel
University of Wollongong
Wollongong, NSW
Australia

ISSN 2191-8112   ISSN 2191-8120   (electronic)
ISBN 978-1-4614-4701-6   ISBN 978-1-4614-4702-3   (eBook)
DOI 10.1007/978-1-4614-4702-3
Springer New York Heidelberg Dordrecht London

Library of Congress Control Number: 2012940952

Printed on acid-free paper

Springer is part of Springer Science+Business Media (www.springer.com)

Contents

Chapter 1
Introduction

With the seemingly infinite amount of information available in online environments, a growing number of users seeks an effective way to find information online. Accordingly, recommender systems that provide personalized support to online users in their information search and decision-making are increasingly seen as necessary and critical components of the online user's web experience (Ochi et al. 2010; Zanker and Ninaus 2010). Recommender systems are available across various domains, including online dating, travel, books, movies, electronics, etc. Yet, although these systems are expected to support online users in complex decision-making processes, they are often not used efficiently due to a lack of confidence in the recommendations they provide (Moulin et al. 2002). Recent survey findings (ChoiceStream 2009) indicated that more than one-half (59 %) of Internet users were not happy with the product recommendations they received at e-commerce sites. These findings suggest that it is important for recommender system research to examine factors that influence the likelihood of recommendations to be accepted and integrated into decision-making processes. Most recommender system research has focused on improving the matching algorithms while a considerably smaller stream of research has explored factors that influence qualities of the system-user interaction (Mahmood et al. 2008). Interactions with recommender systems are in essence conversations that should be examined from a communication point of view (Lucente 2000). The traditional persuasion literature suggests that people are more likely to accept recommendations when the sources display persuasive cues during the interaction process. Recommender systems are sources with the need to persuade their users. Indeed, it has been argued that creating a persuasive recommender system is important in increasing the likelihood of recommendation acceptance (Fogg 2003; Dijkstra et al. 1998; Jiang et al. 2000; Zanker et al. 2006; Gretzel and Fesenmaier 2007; Nguyen et al. 2007; Yoo and Gretzel 2008). The question of how to actually translate persuasiveness into system characteristics in the context of recommender systems, however, still underexplored.

Existing research conducted from a communication perspective suggests that technologies can be more persuasive when leveraging social aspects that elicit

K.-H. Yoo et al., *Persuasive Recommender Systems*, SpringerBriefs in Electrical and Computer Engineering, DOI: 10.1007/978-1-4614-4702-3_1, © The Author(s) 2013

social responses from their human users (Fogg 2003; Nass and Moon 2000). This notion emphasizes the role of recommender systems as quasi-social actors who interact with users socially. If seen as a social communication process, it becomes clear that the characteristics of recommender systems displayed to users in the interaction process influence the perceptions of their users. Various factors have been investigated in the traditional persuasion literature based on human–human communication. Recent studies in the context of human–computer interaction found that these characteristics are also important when humans interact with technologies (Fogg 2003; Fogg et al. 2002; Nass and Moon 2000; Reeves and Nass 1996). With regards to recommender systems, some studies have empirically investigated the persuasive role of recommender systems (e.g. Cosley et al. 2003; Gretzel and Fesenmaier 2007; McNee et al. 2003; Nguyen et al. 2007; Pu and Chen 2007; Qiu 2006; Yoo 2010; Zanker et al. 2006). While these studies identified a number of important factors that help to develop more persuasive recommender systems, still many other factors have not been examined. Further, relevant findings are scattered across the academic literatures in computer science, marketing, management, communication and so on. Thus, the existing recommender system literature does not provide a comprehensive framework for understanding recommender systems as persuasive advice givers.

This book therefore seeks to integrate existing insights into a conceptual framework for persuasive recommender systems. For this purpose, Chap. 2 first provides the theoretical background for understanding recommender systems as persuasive social actors. The factors identified in traditional persuasion literature are briefly reviewed in Chaps. 3–5. In addition to providing overviews, the chapters discuss how these factors have been studied in technology contexts and, in particular, in the recommender systems realm. Chapter 6 offers a summary and further discussion. Finally, implications for recommender system design are presented in Chap. 7 and Chap. 8 suggests directions for future research based on identified research gaps. Overall, by exploring existing findings and identifying important knowledge gaps, this book seeks to provide insights for recommender system researchers as far as future research needs are concerned. It also aims at providing practical implications for recommender system designers who seek to enhance the persuasive power of the recommender systems they build.

Chapter 2
Theoretical Background

2.1 Relevant Communication Theories

The conceptual framework and the underlying principles for persuasive recommender systems are developed based on theoretical background emerging from two theoretical streams in communication research: the Communication-Persuasion Paradigm and Media Equation Theory.

2.1.1 Communication-Persuasion Paradigm

A recommendation is persuasive when it results in attitude or behavior change. The ultimate measure of success for a recommender system is of course actual choice of the recommended alternative. The extent to which a recommendation influences its receiver depends on (1) its form and content; (2) its source; (3) its receiver and his/her characteristics and (4) contextual factors (O'Keefe 2002). These factors are fundamental components of the communication-persuasion paradigm and are interrelated with each other in persuasion processes (Michener et al. 2004). Figure 2.1 displays this paradigm and shows how these elements are interrelated.

As illustrated in this figure, the persuasive outcomes are influenced by multiple factors within each component. First, the characteristics of the source can influence how the message is perceived by the message receiver. Second, the variables related to the message itself can significantly influence its persuasiveness by determining the way it is processed. Third, the experience or characteristics of the message receiver play a role when he/she decides whether to accept the message or not. Finally, a number of contextual factors can affect the persuasive outcomes throughout the process.

Not surprisingly, a great number of studies have identified these factors and systematically tested their influence on persuasive outcomes (O'Keefe 2002).

K.-H. Yoo et al., *Persuasive Recommender Systems*, SpringerBriefs in Electrical and Computer Engineering, DOI: 10.1007/978-1-4614-4702-3_2, © The Author(s) 2013

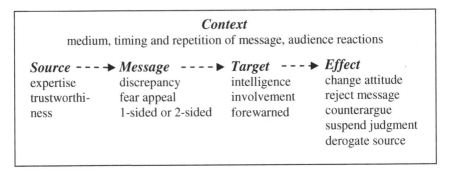

Fig. 2.1 The communication-persuasion paradigm (Michener et al. 2004; O'Keefe 2002)

For example, numerous empirical investigations have found that a communicator's message is more persuasive when the communicator is perceived as credible and likeable by the message receiver (Andersen and Clevenger 1963; Atkin and Block 1983; Baker and Churchill 1977; Friedman and Friedman 1979; Hovland and Weiss 1951; Kelman and Hovland 1953; Patzer 1983). Many studies also found that more specific recommendations are more persuasive than general recommendations (Evans et al. 1970; Frantz 1994; Leventhal et al. 1966; O'Keefe 1997). Message receivers' involvement with the issue (Johnson and Eagly 1989; Petty and Cacioppo 1990) and their intelligence (Rhodes and Wood 1992) are also found to be influential in persuasion processes. The factors investigated in past studies are discussed in greater detail in the following chapters.

These persuasive factors identified in traditional communication research have recently been tested in technology-mediated communication contexts and have been found to be equally important when people communicate using technologies. Flanagin and Metzger (2003) noted that it is possible to translate several components of source credibility to the online environment. For example, they suggested that expertise may be communicated through the accuracy and comprehensiveness of a Web site's information, its professionalism and its sponsor's credentials while trustworthiness is associated with a Web site's integrity as demonstrated through its policy statements, use of advertising as well as firm or author reputation. Fogg (2003) also found that source credibility matters when humans interact with computers. In addition, authority cues have been found to enhance online users' credibility judgments of a computing technology (Fogg 2003) and of online reviewers (Yoo et al. 2007). Online users have also been found to be more easily persuaded by technology that is similar to them in some way (Moon 2002; Fogg 2003). Some studies have found that a physically attractive virtual character was more favorably evaluated by users (Fogg 2003) and served as a more effective sales agent (Holzwarth et al. 2006). The findings of these empirical studies indicate that the persuasive cues investigated in human–human communication could be effectively incorporated in technology contexts to make interactions more persuasive.

Fogg (2003) suggested that understanding the persuasive social role of technology is essential especially when computers take the role of instructing or advising computer users. Since the role of recommender systems involves giving advice in online environments, traditional studies of persuasive factors could provide an important framework to examine the interaction between users and systems as well as users' evaluations of systems.

2.1.2 Media Equation Theory

It seems obvious that a recommender system is a tool or medium, not an actor in social life. However, media equation theory suggests that individuals' interactions with computers, television, and new media are fundamentally social and natural, just like people's interactions with other people in real life (Reeves and Nass 1996). According to Reeves and Nass (1996), people unconsciously and auto-matically apply social rules when they interact with media. This theory thus argues that technologies should be understood as social actors, not just as tools or media.

Several empirical studies have supported this notion of computers as social actors. For example, a number of studies (Nass et al. 2000; Nass et al. 1997) has found that people apply gender and ethnicity stereotypes to computers. Nass and his colleagues (1997) found that people evaluated a computer as significantly more competent when it provided tutoring with a male voice rather than a female voice. They also found that the female-voiced computer was rated as a better teacher than a male-voiced computer when the computer discussed love and relationships which is a stereotypically female topic. But, the computer users rated it as a worse teacher when it talked about computers, which is a stereotypically male topic. Other studies (Nass et al. 2000; Qiu 2006) have found that computer users perceived same-ethnicity embodied computer agents as more attractive, trust-worthy, persuasive, and intelligent than different-ethnicity agents. This indicates that similarity rules important in offline contexts also apply when humans interact with computers.

The findings of Fogg and Nass (1997) also revealed that people exhibit social behaviors such as politeness and reciprocity toward computers. In their experi-ment, study participants worked with computers to learn about some facts and then were asked to evaluate the computer they had used. Half of the participants were asked to evaluate the computer's performance using the same computer they had worked with while the other half answered identical questions on a different computer located on the other side of the room. The results showed that partici-pants who answered on the same computer gave significantly more positive responses. This suggests that they showed politeness and reciprocity toward the computers they knew and worked with. In addition, Nass and Moon (2000) found that impacts of authority cues, i.e., information is accepted uncritically when it is given by an authority figure, also occurs when people interact with technologies. They manipulated the labeling of machines to see if the labeling cues influenced individuals' perceptions of the content the machine presented and found that the content presented by a "specialist" machine was evaluated significantly higher in quality than content presented by a machine labeled as a "generalist".

According to Fogg et al. (2002), computers function in three basic ways: as tools, as media, and as social actors. While previous recommender system studies largely focused on systems as tools, recent studies (Qiu 2006; Wang and Benbasat 2005; Gretzel and Fesenmaier 2007; Yoo 2010) have argued that users often socially interact with recommender systems. Thus, the social aspects of recommender systems need to be better understood. Media equation theory provides a good theoretical framework for such research.

2.2 Recommender Systems as Persuasive Social Actors

As discussed above, a growing number of studies emphasize the social aspects of technologies (Fogg 2003; Nass and Moon 2000; Reeves and Nass 1996) and the social role of recommender systems has also been suggested and investigated. Zanker and his colleagues (2006) argued that interactions with recommender systems should not only be seen from a technical perspective but should also be examined from a social and emotional perspective. Wang and Benbasat (2005) found that users perceived human characteristics such as benevolence and integrity from recommender systems and treated systems as social actors. The findings by Aksoy et al. (2006) suggest that the similarity rule is also applied when humans interact with recommender systems. They found that a user is more likely to use a recommender agent when it generates recommendations in a way similar to the user's decision-making process. Morkes et al. (1999) demonstrated that computer agents that use humor are rated as more likable, competent, and cooperative. In addition, trust in recommender systems has also been found to be important to support system users' decision-making (Bauernfeind and Zins 2006) as well as intentions to adopt the recommender systems (Wang and Benbasat 2005, 2008). In addition, Gretzel (2004) revealed that the interaction process between users and recommender systems significantly influences users' perceptions of the system and the recommendations provided by such systems. More recently, Yoo (2010) investigated how embedded virtual agents on system interfaces influence users when they evaluate systems. The study found that users socially interact with the systems and the social cues portrayed by the embedded virtual agents influence system users' evaluations of the agents as well as the overall system quality.

These studies all support the notion of recommender systems as social actors and suggest a need for examining the social aspects of recommender systems. This implies that recommender systems can be understood as communication sources to which theories developed for human–human communication apply. Applying such theories opens up a new avenue for understanding the role of recommender systems and their interactions with users.

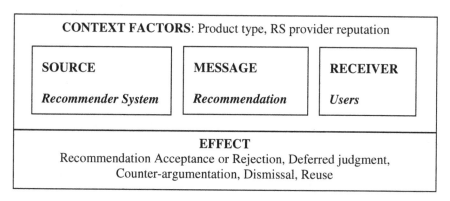

Fig. 2.2 Conceptual framework for persuasive recommender systems

2.3 Conceptual Framework

Applying communication theories to recommender systems, the system itself can be seen as a source, its recommendations as messages and its users as receivers of these messages. These process components exist within a certain communication context that influences the way cues are communicated and perceived. The interaction results in communication effects that ultimately encourage or discourage reuse of the system (Fig. 2.2).

In the following three chapters, the specific persuasive factors (source characteristics, message variation and receiver/context factors) found in human and human communications are reviewed and the chapters discuss how the factors have been adopted and examined in technology contexts, particularly in recommender systems. While there are numerous persuasive factors that have been identified in traditional persuasion literature, the review presented in this book is focused on the characteristics relevant to the recommender system context.

Chapter 3
Source Factors

This chapter reviews the source factors in human–human communication and discusses how the source characteristics have been applied and examined in technology as well as recommender system contexts.

3.1 Source Factors in Human–Human Communication

Hoveland and his colleagues (1953) argued that one of the main classes of stimuli that determine the success of persuasive attempts can be summarized as the observable characteristics of the perceived message source. Naturally, considerable research attention has been focused on investigating the various communicator characteristics that influence the outcomes of the communicator's persuasive efforts in human–human interactions. An overview of the most relevant source factors mentioned in the literature is provided in Fig. 3.1.

3.1.1 Credibility

A good number of past studies have confirmed that a more credible source is preferred and also more persuasive (Anderson and Clevenger 1963; Gilly et al. 1998; Harmon and Coney 1982; Lascu et al. 1995; McGuire 1968; Sénécal and Nantel 2003, 2004). Credibility is generally described as comprising multiple dimensions (Buller and Burgoon 1996; Gatignon and Robertson 1991; Petty and Cacioppo 1981; Self 1996) but most researchers agree that it consists of two key elements: expertise and trustworthiness (Fogg 2003; Fogg et al. 2002; O'Keefe 2002; Rhoads and Cialdini 2002). The dimension of expertise captures the perceived knowledge and skill of the source (Mayer et al. 1995; O'Keefe 2002) while trustworthiness of a source refers to aspects such as character or personal integrity (O'Keefe 2002). Whether a source is perceived as having expertise and being trustworthy depends to a great extent on its characteristics.

K.-H. Yoo et al., *Persuasive Recommender Systems*, SpringerBriefs in Electrical and Computer Engineering, DOI: 10.1007/978-1-4614-4702-3_3, © The Author(s) 2013

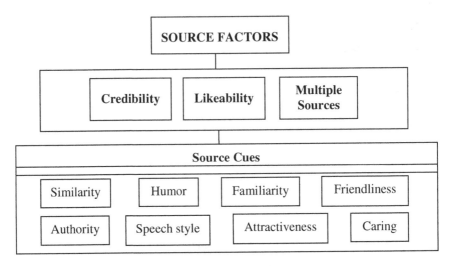

Fig. 3.1 Overview of relevant source factors

3.1.2 Likeability

People mindlessly tend to agree with those who are seen as likable (Burgoon et al. 2002). Liking refers to the affective bond that an individual may feel toward another person (Smith et al. 2005). Research generally supports the assumption that liked communicators are more effective influence agents than are disliked communicators (Eagly and Chaiken 1975; Giffen and Ehrlich 1963; Sampson and Insko 1964) and likability has been labeled a persuasion tactic and a scheme of self-presentation (Cialdini 1994). O'Keefe (2002) stressed enhanced liking for the source is commonly accompanied by enhanced judgments of the communicator's trustworthiness. Further, a number of studies found that similarity increases likeability (Byrne 1971; Carli et al. 1991; Hogg et al. 1993).

There is also some evidence indicating that the receiver's liking of the communicator can influence judgments of the communicator's trustworthiness, although not judgments of the communicator's expertise (O'Keefe 2002; Levine 2003). Similarity cues as well as humor have also been found to increase liking of the source.

3.1.3 Multiple Sources

Social impact theory (Jackson 1987; Latané 1981) explains that impact of a persuasive attempt depends on strength, immediacy and number of influencing sources. The theory predicts that the message will be more persuasive when it comes from multiple sources than from a single source. This prediction was supported by several studies that found that a message presented by several different sources was more persuasive than the same message presented by a single source (Harkins and Patty 1981, 1987; Wolf and Bugaj 1990).

3.1.4 Source Cues

3.1.4.1 Similarity

It is unquestionably the case that perceived similarities or dissimilarities between source and audience can influence the audience's judgment (O'Keefe 2002). In general, homophily theory (Lazarsfeld and Merton 1954) states that humans like similar others. However, the relation between similarity and the dimensions of credibility appears to be complex.

Past empirical studies show contradicting results with respect to similarity and source expertise judgments. For example, Mills and Kimble (1973) found that similar others are seen as having greater expertise than dissimilar others. However, Delia (1975) observed that similarity between the source and the message receiver makes the receiver see the source less as an expert. In contrast, some studies found that similarity does not make any difference in source expertise judgments (e.g., Swartz 1984; Atkinson et al. 1985).

The perceived similarity of the message source also has varying effects on perceived trustworthiness of the communicator. O'Keefe (2002) suggested that perceived attitudinal similarities can influence the receivers' liking for the source, and enhanced liking for the source is commonly accompanied by enhanced judgments of the communicator's trustworthiness. However, Atkinson et al. (1985) found that ethnic similarity and dissimilarity did not influence the perceived trustworthiness of the source, while Delia (1975) observed that similarity sometimes diminished trustworthiness perceptions.

Reflecting on the complex nature of the relationship between similarity and judgments of the communicator's credibility, O'Keefe (2002) noted that the effects of perceived similarities on judgments of communicator credibility depend on whether, and how, the receiver perceives these as relevant to the issue at hand. Thus, different types of similarity likely have different effects in different communication contexts.

3.1.4.2 Symbols of Authority

Evidence presented in the persuasion literature indicates that we often embrace the mental shortcut of assuming that people who simply display symbols of authority such as titles, tailors and tone should be listened to (Rhoads and Cialdini 2002; Bickman 1974; Hofling et al. 1966; Giles and Coupland 1991; Pittam 1994). Hofling et al. (1966) found that something simple as the title "Dr." made subjects perceive a source as credible and was surprisingly effective as a compliance-gaining device.

Similarly, a number of studies reported that cues like the communicator's education, occupation, training, and amount of experience influence a message receiver's perceptions of source credibility. For example, Hewgill and Miller (1965) manipulated the occupations of the communicator (Professor vs. High school sophomore) for the same message and found that those subjects who were informed that the message had been written by a professor evaluated both the source and the message as significantly more credible.

Uniforms and well-tailored business suits are another recognized symbol of authority that can influence credibility judgment and bring on mindless compliance (Rhoads and Cialdini 2002; Cialdini 1994). The findings of Bickman (1974) indicate that a person wearing a security guard's uniform who asks strangers to do things could produce significantly more compliance than a person wearing street clothes. Sebastian and Bristow (2008) revealed that formally dressed individuals achieved greater credibility ratings than individuals who dressed informally.

3.1.4.3 Styles of Speech

Several studies exist which suggest that the style of speech can influence speaker credibility judgments. For instance, several studies have demonstrated that communicators can enhance their trustworthiness when they provide both sides of the argument—the pros and the cons—rather than arguing only in their own favor (Eagly Wood and Chaiken 1978; Smith and Hunt 1978). Cooper Bennett and Sukel (1996) suggest that people evaluate the speaker's expertise higher when he/she speaks in complex, difficult-to-understand terms. This indicates that experts may be most persuasive when nonexperts cannot understand the details of what they are saying (Rhoads and Cialdini 2002). Several investigators have found that with increasing numbers of nonfluencies in a speech, speakers are rated significantly lower on expertise judgments (Burgoon et al. 1990; Engstrom 1994; McCroskey and Mehrley 1969; Schliesser 1968) and the speaking rate can also influence credibility judgments, although the evidence for this effect is not as clear as for others (Addington 1971; Gundersen and Hopper 1976; MacLachlan 1982; Lautman and Dean 1983). Also, citing sources of evidence appears to enhance perceptions of the communicator's expertise and trustworthiness (e.g., Fleshler Ilardo and Demoretcky 1974; McCroskey 1970; O'Keefe 1998).

3.1.4.4 Humor

Previous studies found effects of humor when message receivers evaluate the communicator's credibility. However, the specific effects varied across different studies. A number of studies found positive effects of humor on communicator trustworthiness judgments but rarely on judgments of expertise (Chang and Gruner 1981; Gruner and Lampton 1972; Tamborini and Zillmann 1981). When positive effects of humor were found, the effects tended to enhance the audience's liking of the communicator and this liking helped increase perceptions of trustworthiness. In contrast, some researchers found that the use of humor can decrease the audience's liking for the communicator, the perceived trustworthiness, and even the perceived expertise of the source when the use of humor is perceived as excessive or inappropriate for the context (Bryant et al. 1981; Munn and Gruner 1981; Taylor 1974).

3.1.4.5 Physical Attractiveness

A number of studies have found that physically attractive communicators are more persuasive (Horai et al. 1974; Snyder and Rothbart 1971; Eagly et al. 1991). Eagly et al. (1991) explained that there appears to be a positive reaction to good physical appearance that generalizes to favorable trait perceptions such as a talent, kindness, honesty and intelligence. The effects of physical attractiveness are seen as influencing indirectly, especially by means of influence on the receiver's liking for the communicator (O'Keefe 2002).

3.1.4.6 Caring

Caring as a theoretical construct encompasses motives and intentions. Benevolence, which refers to concern about the message receiver's best interest, has been proposed as an underlying dimension of trust (Bart et al. 2005). Delgado-Ballester (2004) also conceptualizes good intentions as an important factor that determines trustworthiness. Perloff (2003) reports that communicators who have the recipient's interests at their heart communicate goodwill, which is a core aspect of credibility.

3.1.4.7 Familiarity

As a rule, individuals are more likely to comply with requests of someone they know in contrast to those made by strangers (Cialdini 1993). Cialdini (1993) reports that for this effect to work the known person does not even have to be present, sometimes dropping a name can be enough. Also, we are more prone to like people we know personally (Levine 2003; Cialdini 1994). In addition, it is a fact of social interaction that people are more favorably inclined toward the needs of those they know (Shavitt and Brock 1994).

3.1.4.8 Friendliness

Praise and other forms of positive estimation stimulate liking (Byrne and Rhamey 1965). Communicators who are nice can change attitudes because they make the recipient feel good, and the positive feeling becomes transferred to the message (Rhoads and Cialdini 2002). For instance, research indicates that drawing a happy, smiling face on the back of checks before giving them to customers increases tip size (Perloff 2003).

3.2 Applying Source Factors to Technology

3.2.1 Source Factors in Technology Contexts

Many recent studies have investigated how certain characteristics of technologies influence their users' perceptions and behaviors. Similarity between a computer and its users was found to be important when computer users evaluated the computer and its contents (Nass and Moon 2000; Fogg 2003). For example, Nass and Moon (2000) report that computers that convey similar personality types are more persuasive. In their study, dominant participants were more attracted to, assigned greater intelligence to, and conformed more with a dominant computer compared to a submissive computer. Similarly, submissive participants significantly more positive reactions to the submissive computer as opposed to the dominant computer, despite the essentially identical content displayed by both types of computers. Nass et al. (2000) also revealed effects of demographic similarity. Their study found that computer users perceived computer agents as more attractive, trustworthy, persuasive and intelligent when same-ethnicity agents were presented.

Presenting authority symbols has also been identified as an influential factor when people interact with technology. Nass and Moon (2000) found that a television set labeled as a specialist was perceived as providing better content than a television set labeled as a generalist. Fogg (2003) also posited that computing technology that assumes roles of authority is more persuasive. He argued that websites displaying awards or third-party endorsements such as seals of approval will be perceived as more credible.

A number of studies (Nass et al. 1997; Nass et al. 2000) argue that the demographic characteristics of computer agents influence users' perceptions. As discussed earlier, Nass Moon and Green (1997) illustrated that people apply gender and ethnicity stereotypes to computers. Their study found that people evaluated the tutor computer as significantly more competent and likeable when it was equipped with a male voice than a female voice. They also found that the female-voiced computer was perceived as a better teacher of love and relationships and a worse teacher of computing than a male-voiced computer, even though they performed identically.

In addition, the use of language such as flattery (Fogg and Nass 1997), apology (Tzeng 2004) and politeness (Mayer et al. 2006) has been identified as a factor which makes a difference in computer users' perceptions and behaviors. Further, the physical attractiveness of computer agents was found to matter. The findings by Nass et al. (2000) indicate that computer users prefer to look at and interact with computer agents that are more attractive.

Finally, humor has also been tested in the human–computer interaction context. Morkes et al. (1999) found that computers which display humor are rated as more likeable. Further, Kang and Gretzel (2012) demonstrated multiple source effects in the context of podcast audio tours in a national park.

3.2.2 Source Factors in Recommender Systems

A number of previous studies have investigated how specific characteristics of recommender systems influence users' evaluations of the system as well as its recommendations. Existing recommender system studies have examined some source factors identified as influential in traditional interpersonal relations and also identified important source factors that are prominent in recommender system contexts. Xiao and Benbasat (2007) classified the various source characteristics that have been studied as being associated with either recommender system type, input, process or output design. Also with the increasing interest in and use of embodied agents in recommender systems, a growing number of studies has investigated the effects of characteristics displayed by embodied virtual agents that often guide users through the various steps of the recommender process.

3.2.2.1 Recommender System Type

Recommender systems come in different shapes and forms and can be classified based on filtering methods, decision strategies or amount of support provided by the recommender systems for consumer purchases (Xiao and Benbasat 2007). Several previous studies have discussed the advantages and disadvantages of these different types of recommender systems (e.g. Ansari et al. 2000; Maes et al. 1999; Burke 2002). Different filtering methods were compared and it was found that meta-recommender systems that combine collaborative filtering and content filtering are evaluated as more helpful than traditional systems that use a pure collaborative filtering technique (Schafer et al. 2002, 2004). Burke (2002) also confirmed that hybrid recommender systems provide more accurate predictions of user preferences. Regarding the different decision strategies used in recommender systems, compensatory recommender systems have been suggested to lead to greater trust, perceived usefulness and satisfaction than non-compensatory recommender systems (Xiao and Benbasat 2007). They have also been found to increase users' confidence in their product choices (Fasolo et al. 2005).

As far as the amount of support provided by recommender system is concerned, Xiao and Benbasat (2007) argued that needs-based systems rather than feature-based systems help users better recognize their needs and more accurately answer the preference-elicitation questions, thus resulting in better decision quality. Needs-based systems are therefore recommended for novice users (Felix et al. 2001).

Some studies found that conversational recommender systems can be more persuasive (Lucente 2000; Zanker et al. 2006) since the systems can cope with the natural language input of the customer and allow for adaption of the process (Mahmood et al. 2008) while other types of recommender systems only allow pre-structured interaction paths.

3.2.2.2 Input Characteristics

Input characteristics of recommender systems include those cues that are related with the preference elicitation method, ease of generating new/additional recommendations and the amount of control users have when interacting with the recommender systems' preference elicitation interface (Xiao and Benbasat 2007). A number of previous findings suggest that characteristics associated with recommender system input design influence system users' evaluations. Xiao and Benbasat (2007) specifically argued that the preference elicitation method (implicit vs. explicit) influences users' evaluation of the system. They proposed that an implicit preference elicitation method leads to greater perceived ease of use of and satisfaction with the recommender system while explicit elicitation is considered to be more transparent for users and leads to better decision quality.

Allowing users more control was also found to be an influential factor when evaluating systems. West et al. (1999) posited that giving more control to system users will increase their trust and satisfaction with the system. Indeed, a study conducted by McNee et al. (2003) found that users who used user-controlled interfaces reported higher user satisfaction than users who interacted with system-controlled and mixed-initiative recommender systems. In addition, users of user-controlled interfaces felt that the recommender systems more accurately represented their tastes and showed the greatest loyalty to the systems. Similarly, Pereira (2000) demonstrated that users showed more positive affective reactions to recommender systems when they had increased control over the interaction with the recommender system. Komiak et al. (2005) also found that control over the process was one of the top contributors to users' trust in a virtual agent. Supporting the importance of user control, Wang (2005) noted that more restrictive recommender systems were considered as less trustworthy and useful by their users.

In addition to control, the structural characteristics of the preference elicitation process (relevance, transparency and effort) have also been found to influence users' perceptions of the recommender system (Gretzel and Fesenmaier 2007). The specific study by Gretzel and Fesenmaier (2007) found that topic relevance, transparency in the elicitation process and the effort required by users to provide inputs positively influence users' perceptions of the value of the elicitation process. The findings suggest that by asking questions, the system takes on a social role and communicates interest in the user's preferences, which is seen as valuable. The more questions it asks, the greater its potential to provide valuable feedback. Also, making intentions explicit in this interaction is important. Although trust was not specifically measured, benevolence and intentions are important drivers of trust and can be implied from the importance based on transparency. Further, McGinty and Smyth (2002) suggested that the conversation style of recommender systems during the input process matters. In contrast to Gretzel and Fesenmaier (2007), they argued that the comparison-based recommendation approach which asks users to choose a preferred item from a list of recommended items instead of a current deep dialogue approach that asks users a series of direct questions about the importance of product features would minimize the cost to the user and maintain recommendation quality.

3.2.2.3 Process Characteristics

Characteristics of recommender systems displayed during the recommendation calculation process appear to influence users' perceptions of the systems (Xiao and Benbasat 2007). Such process factors include information about the search process and about the system response time. Mohr and Bitner (1995) noted that system users use various cues or indicators to assess the amount of effort saved by decision aids. Indicators that inform users about the search progress help users become aware of the efforts saved by the system. The higher users' perceptions of the effort saved by decision aids the greater their satisfaction with the decision process (Bechwati and Xia 2003). Sutcliffe et al. (2000) found that users reported usability/comprehension problems with information retrieval systems that did not provide a search progress indicator.

Influences of system response time, i.e. the time between the user's input and the system's response, have also been identified as important in a number of studies. Basartan (2001) varied the response time from a simulated shopbot and found that users prefer those shopbots less that make them wait a long time before receiving recommendations. In contrast, Swearingen and Sinha (2001, 2002) found that the time taken by users to register and to receive recommendations from recommender systems did not have a significant effect on users' perceptions of the system. In the study by McNee et al. (2003), the lengthier sign up process increased users' satisfaction with and loyalty toward the system. Xiao and Benbasat (2007) explained that the contradicting findings of previous studies regarding response time may depend on users' cost-benefit assessments. They suggest that users do not form negative evaluations of the recommender systems when they perceive the benefits of waiting as leading to high quality recommendations. The findings of Gretzel and Fesenmaier (2007) regarding the relationship between elicitation effort and the perceived value of the elicitation process support this assumption.

Providing playfulness features during interactions has also been suggested as an important persuasion factor. Kim and Morosan (2006) argued that recommender systems can be more credible and persuasive by integrating playful features. They explained that the playfulness of recommender systems can enhance users' experiences online by allowing them to fully immerse themselves into the online experience.

3.2.2.4 Characteristics of Embodied Agents

Recommender systems often include virtual personas guiding the user through the process. It can be assumed that social responses are even more prevalent if the system is personified. Indeed, the important role and impacts of embodied interface agents in the context of recommender systems have recently been emphasized in a number of studies. For example, the presence of a humanoid virtual agent in the system interface was found to increase system credibility (Moundridou and Virvou 2002), to augment social interactions (Qiu 2006), to stimulate user involvement (Zanker et al. 2006), to enhance the online shopping experience (Holzwarth et al. 2006), as well as to induce trust (Wang and Emurian 2005). With growing interests in such interface agents, a number of studies have started investigating if and how certain characteristics of the interface agent influence recommender system users' perceptions and evaluations.

One of the important identified characteristics of agents is anthropomorphism. Many researchers have found that anthropomorphism of embodied agents influences people's interactions with computers (e.g. Koda 1996; Nowak and Biocca 2003; Nowak 2004) and specifically with recommender systems (Qiu 2006; Yoo 2010). Yet, the benefits and costs of anthropomorphic agents are debatable. For example, more anthropomorphic interface agents were rated as being more credible, engaging, attractive and likeable than less anthropomorphic agents in some studies (Koda 1996; Nowak and Rauh 2005) while other studies found contrasting results (Nowak 2004; Nowak and Biocca 2003; Murano 2003). The social cues communicated by the inclusion of such agents might create expectations in the users that cannot be met by the actual system functionalities.

Human voice is a very strong social cue that has been found to profoundly shape human-technology interactions (Nass and Brave 2005). However, findings in the context of embodied interface agents are not widely available and are currently inconclusive. The voice output of interface agents was found to be helpful in inducing social and affective responses from users in some studies (Qiu 2006; Moreno et al. 2001; Yoo 2010) but other studies found that sociability was higher when the system avatar only communicated with text (Sproull et al. 1996).

The demographic characteristics of interface agents have also been found to influence system users' perceptions and behaviors. Qiu (2006) reported that system users evaluated the system as more sociable, competent, and enjoyable when the agents were matched with them in terms of ethnicity and gender, thus supporting the homophily hypothesis. Cowell and Stanny (2005) also observed that system users prefer to interact with interface characters that matched their ethnicity and were young looking. A study by Nowak and Rauh (2005) indicated that people showed a clear preference for characters that matched their gender.

In addition to similarity cues, other source characteristics have also been investigated in the context of embodied interface agents. The effects of attractiveness and expertise of interface agents were tested by Holzwarth et al. (2006). They found that an attractive avatar is a more effective sales agent at moderate levels of product involvement while an expert agent is a more effective persuader at high levels of product involvement. Further, the potential impacts of non verbal behavior cues including facial expression, eye contact, gestures, para-language and posture of interface agents were emphasized by Cowell and Stanney (2005). Recent experiments (Yoo 2010) empirically investigated how virtual agents' anthropomorphic, authority and similarity cues influence recommender system users' perceptions. The findings indicate that the human-like agent was perceived as more attractive than the object agent and liking toward the system was increased when the system presented a virtual agent with voice than one without voice. The study further examined the impacts of agents' outfit as well as age. System users evaluated the virtual agent as significantly higher in expertise when it wore a uniform as compared to a casual outfit. They also thought an agent similar in age to theirs was more attractive than an older agent. However, research in this area is still somewhat limited. Especially, with the rapid evolvement of the online virtual technology field, additional influential interface agent characteristics that will emerge should be identified and examined.

Chapter 4
Message Factors

While the source of message plays an important role in the persuasion process, the message itself can also have a significant impact on its persuasiveness (Michener et al. 2004; O'Keefe 2002). This chapter briefly reviews the message factors studied in persuasion literature and discusses how those factors have been applied and examined in the recommender system realm.

4.1 Message Factors in Human–Human Communication

Messages differ in their contents, structure as well as the way they are presented to targets. O'Keefe (2002) illustrated that the message factors discussed in past studies fall under three categories: message structure, message content and sequential-request strategies (Fig. 4.1).

4.1.1 Message Structure

Extensive research has been conducted on how the structure of a message can influence its persuasiveness, including *order of presentation, conclusion drawing, message specificity and message format.*

Previous research generally indicates that the arguments presented first and last are recalled better than those presented in the middle (Krugman 1962; O'Keefe 2002). This suggests that a communicator's important arguments should be presented early or late in the message but not in the middle. However, many other studies noted that varying the order makes little difference to overall persuasive effectiveness (Gilkinson et al. 1954; Gulley and Berlo 1956). Thus, the arrangement of arguments in a message needs to be sensitive to the particulars of the persuasion circumstances. There is an indication that the effects of presentation order can vary depending on the message receiver's elaboration. Haugtvedt and

K.-H. Yoo et al., *Persuasive Recommender Systems*, SpringerBriefs in Electrical and Computer Engineering, DOI: 10.1007/978-1-4614-4702-3_4, © The Author(s) 2013

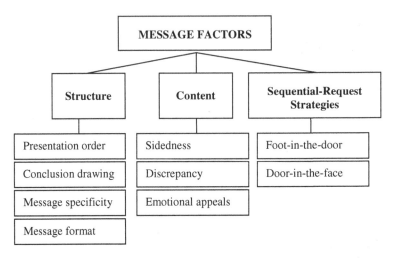

Fig. 4.1 Overview of relevant message factors

Wegener (1994) suggested that a message presented first can be more persuasive when the message receiver's elaboration is high while a message presented last tends to be more effective when elaboration is low. The study explains that the first message can produce targeted attitudes since highly motivated message receivers perceive that the message is interesting and familiar. In contrast, the last message can be more persuasive when audiences are less involved because the last message is more prominent in their memories.

Researchers have also examined whether a message should explicitly state a firm conclusion or let receivers figure the conclusion out themselves. The research evidence suggests, in general, messages containing explicit conclusions are more effective than messages that omit such statements (O'Keefe 1997; Struckman-Johnson and Struckman-Johnson 1996). However, some studies have shown that the effectiveness of conclusion drawing may depend on the message receivers, the type of topic, and a variety of situational factors (Hovland and Mandell 1952). For example, Chance (1975) found that open-ended advertising messages were more memorable and effective for some cases since the unanswered conclusion allows message receivers to draw their own conclusions and therefore reinforces the points being made in the message. This suggests that the target audiences' characteristics should be considered when a communicator designs the message structure.

Another message structure decision involves message specificity. Past studies compared messages that provide a general description of the advocate's recommended action to the messages that provide a more detailed recommendation. The findings of such studies indicate that more specific recommendations are more persuasive than general recommendations (Evans et al. 1970; Frantz 1994; O'Keefe 1997).

In addition, the impacts of message format have been examined in some studies. Previous studies have compared the persuasiveness of messages presented in different formats (written, audiotaped, and videotaped messages) and suggest that there is no general advantage associated with one or another of these forms (Sparks et al. 1998; Pfau et al. 2000; Wilson and Sherrell 1993). However, the findings suggest that the communicator's characteristics take a greater role in influencing persuasive outcomes when the message is presented in audiotape or videotape formats. It was explained that audiotape and videotape formats provide more information about the communicator than the written format, thus enabling the message receiver to decode and evaluate source characteristics more effectively. Alternatively, messages in the written format are more likely to enhance the impact of message content variation and dampen the influence of communicator characteristics (O'Keefe 2002).

4.1.2 Message Content

Previous studies have tested different content variables such as *message sidedness, discrepancy and emotional appeals* to investigate their persuasive effects.

Message sidedness has been examined in a number of studies. Researchers compared a one-sided message that mentions only supporting arguments with a two-sided message that presents both supporting and opposing arguments. The findings show no general difference in persuasiveness between one-sided and two-sided messages but rather there appear to be many possible moderating factors (O'Keefe 2002). One of the moderating factors is the nature of the target audience (Michener et al. 2004). Studies found that one-sided messages work better when the target audiences already agree with the source and they don't know much about the issue while two sided messages are more effective when the target audiences hold the opposing opinions or know a lot about the alternative positions (Karlins and Abelson 1970; Pechmann 1992; Sawyer 1973).

A number of investigations have examined how the variations in message discrepancy—the difference between the position advocated by the message and the target audience's position—influence persuasive outcomes. While there is no simple answer for the relationships between message discrepancy and persuasive outcomes, an inverted U-shaped curve reasonably explains the relationship. This suggests that the messages that are moderately discrepant are more effective in changing a target's opinion and attitudes than messages that are only slightly or extremely discrepant. However, the effects of message discrepancy have been found to be influenced by a number of factors including source credibility and message receiver's involvement. For example, message receivers are more likely to accept a highly discrepant message from a highly credible source than from a less credible source (Aronson et al. 1963; Fink et al. 1983).

In addition, emotional appeals to fear or humor have been found to be the effective techniques (Belch and Belch 2009; O'Keefe 2002) in persuasion attempts.

4.1.3 Sequential-Request Strategies

The effectiveness of two sequential-request strategies was investigated in some previous studies. One strategy is *the foot-in-the-door (FITD) strategy* that initially makes a small request, and then makes the larger target request. In contrast, *the door-in-the-face (DITF) strategy* begins from a large request, which the receiver turns down, and then makes the smaller target request. Previous studies have shown that target audiences are more likely to accept requests when these strategies were used compared to only asking the target request in the first place (Cialdini et al. 1975; DeJong 1979; Freedman and Fraser 1966; O'Keefe 2002).

4.2 Applying Message Factors to Technology

Previous findings indicate that the content and format of recommendations can have a significant impact on a recommender system user's evaluation of a system as well as the recommendation itself (e.g. Cosley et al. 2003; Sinha and Swearingen 2001; Xiao and Benbasat 2007; Wang and Benbasat 2007). In these recommender system studies, the influences of message discrepancy, specificity, sidedness and presentation format (text vs. visual) have been tested. In addition, the importance of transparent explanations, recommendation display layout and site navigation were also investigated.

4.2.1 Recommendation Content

In a recommender system context, it has been found that the recommendations that are only slightly discrepant from system users' positions are more persuasive than highly discrepant recommendations. Swearingen and Sinha (2001) found that recommended products that were familiar to users were helpful in establishing users' trust toward recommender systems. A study by Cooke et al. (2002) also observed that unfamiliar recommendations lowered users' evaluations of recommender systems. While these findings are not consistent with the results in traditional persuasion literature that suggests maximum effectiveness of messages with moderate levels of discrepancy, it may indicate that recommender systems are still not perceived as a highly credible source of advice. Past studies have found that message receivers are more likely to accept a highly discrepant message from a highly credible source but not from a source perceived as low in credibility. This suggests that the influence of message discrepancy should be further investigated as the system technology evolves as well as an increasing number of people use and get familiar with recommender systems.

More specific recommendations appeared to positively influence users' perceptions of recommender systems. Sinha and Swearingen (2001) suggest that detailed product information available on the recommendation page enhances users' trust in the recommender system. Cooke et al. (2002) also explained that the attractiveness of unfamiliar recommendations can be increased if recommender systems provide detailed information about the new product. Similarly, Gretzel (2006) argued that integrating narrative descriptions in recommendations can help the systems better match various users' preferences and also provide system users with means to effectively process the recommended information. A recent empirical finding (Ozok et al. 2010) also supports users' preference for specific recommendations.

In addition, a considerable number of studies examined that explaining why certain items were suggested is important to enhance users' trust toward the systems. Wang and Benbasat (2007) found that explanations of the recommender system's reasoning logic strengthened users' beliefs in the recommender system's competence and benevolence. Herlocker et al. (2000) also reported that explanations were important in establishing trust in systems since users were less likely to trust recommendations when they did not understand why certain items were recommended to them. Bonhard and Sasse (2005) emphasized that recommender systems must establish a connection between the advice seeker and the system through explanation interfaces in order to enhance the user's level of trust in the system. Similarly, Zanker and Ninaus (2010) explained that recommender system's perceived usefulness is enhanced when the system provides informative explanations about why a certain item was recommended. Additional studies (Pu and Chen 2007; Tintarev and Masthoff 2007) also confirmed that system users exhibited more trust in the case of explanations integrated in the interfaces.

The influence of message sidedness was also tested. Nguyen and his colleagues (2007) compared one sided recommendations with two-sided recommendations and found that system users perceived that two-sided messages were significantly easier to follow, less boring and more persuasive.

4.2.2 Recommendation Format

The format in which recommendations are presented to the user also appears to influence users' evaluation of recommender systems. Recommendations were found to be more persuasive when recommender systems presented them using both text and video in contrast to text and image combinations or text only formats (Nanou et al. 2010). It seems that users are more likely to accept rich multimedia recommendations since users can use more information when they evaluate the recommended items.

The interface navigation and layout of the page presenting the recommendation was found to be a significant factor determining users' satisfaction with the system (Sinha and Swearingen 2001; Swearingen and Sinha 2001). For example, Shinha and Swearingen (2001) found that users were generally dissatisfied when they needed to execute too many clicks to access the item information or if only a few recommendations were displayed on each screen. Consistent with these findings, Yoon and Lee (2004) showed that interface design and display format influenced system users' behaviors. A recent empirical study (Ozok et al. 2010) suggested the recommendations should be placed on the lower-middle section of the screen and the recommended items should not amount to more than three items per main product screen. However, a study conducted by Bharti and Chaudhury (2004) did not find any significant influence of navigational efficiency on users' satisfaction. In addition, Schafer (2005) suggested that merging the preferences interface and the recommendation elicitation interface within a single interface can make the recommender system be seen as more helpful since this "dynamic query" interface can provide immediate feedback regarding the effect caused by the individual's preference changes. Since such an approach merges the input with the output interface, this suggestion touches upon cues such as transparency already discussed in the context of source characteristics.

Chapter 5
Receiver and Context Factors

Although a communicator possesses persuasive characteristics and delivers a persuasive message to a target audience, it is not guaranteed that the message will be accepted by the receiver. The characteristics of a receiver as well as context factors often moderate the persuasion process. This chapter discusses important factors related to the receiver and the context. The factors examined in traditional interpersonal communication studies are reviewed first and the recommender system users' characteristics as well as contextual cues tested in existing recommender system studies are presented.

5.1 Receiver Factors in Human–Human Communication

A number of influential receiver cues have been discussed in past studies. The identified receiver factors can be categorized as natural receiver characteristics such as sex and personality traits and induced receiver factors (Fig. 5.1).

5.1.1 Natural Receiver Characteristics

One important natural receiver characteristic that affects persuasion is the degree of involvement in the issue by the receiver (Johnson and Eagly 1989; Petty and Cacioppo 1990). As explained in the Elaboration Likelihood Model (Petty and Cacioppo 1986), relevancy of the issue affects the way a message receiver processes a message. The more personally relevant the message is, the more involved the receiver will become involved in its processing. Therefore, for a relevant topic, a target more carefully evaluates the message whereas he or she often decides based on peripheral cues when the relevancy level is low. Peripheral processing is low elaboration processing and therefore leads to less stable attitudes, which are more easily attackable.

K.-H. Yoo et al., *Persuasive Recommender Systems*, SpringerBriefs in Electrical and Computer Engineering, DOI: 10.1007/978-1-4614-4702-3_5, © The Author(s) 2013

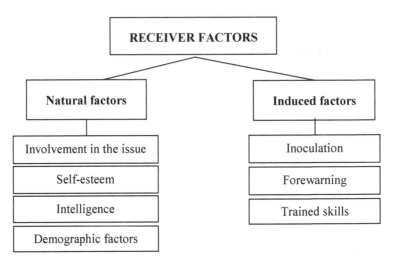

Fig. 5.1   Overview of relevant receiver factors

Message receivers' self-esteem and intelligence also appear to be influential in the persuasion process. The previous findings indicate that the targets are more likely to be persuaded when they possess intermediate levels of self-esteem and lower levels of intelligence (Rhodes and Wood 1992). It was explained that a person with low self-esteem is unlikely to pay sufficient attention to the message and those with high self-esteem are likely to be confident in their current beliefs. As far as intelligence is concerned, a person with greater intelligence is more likely to critically scrutinize messages.

In addition to these factors, the effects of gender (Becker 1986; Eagly and Carli 1981), personality (Zuckerman 1979), age (Omoto et al. 2000) and cultural background (Han and Shavitt 1994) on persuasive outcomes have been discussed. However, these studies often yielded complex results and do not allow for definite conclusions.

5.1.2 Induced Receiver Factors

While message receivers' natural enduring states or characteristics can influence persuasion processes, previous studies suggest that induced recipient characteristics also play a role in persuasion.

In a persuasion process, convincing someone to take on one's point of view is the ultimate goal, but once the target is persuaded, the question is how the viewpoint can persist despite of counter persuasion efforts that the person might encounter. According to inoculation theory (McGuire 1964), an individual can develop resistance to persuasion by being exposed to weak attacks ahead of stronger persuasion attempts. The theory explains that the inoculation builds up the message receiver's resistance and prepares the person to resist stronger attacks on their attitudes in the future. Some research (McGuire 1961; McGuire and Papageorgis 1961) has demonstrated the effectiveness of inoculation treatments to develop a person's ability to resist persuasion attempts.

Another way to induce resistance is simply warning the message receivers that they are about to be exposed to a persuasion attempt. A fair amount of research has examined the forewarning effects in persuasion and suggests that an individual is less likely to change a belief if the person is alerted to the possibility of a belief-attacking message. Forewarning creates a threat which stimulates the production of belief defenses and, thus, will decrease the effectiveness of the attack once it is presented (Petty and Cacioppo 1977; Jackson and Devine 2000). A number of past studies have supported this notion and found that forewarning a message receiver of an impending counter attitudinal message can inhibit opinions change (Dean et al. 1971; McGuire 1966; Papageorgis 1968). Dean and his colleagues (1971) report that forewarning conditions produce less opinion change, regardless of issue involvement and source status. However, other studies have noted possible inter-action effects between warning and topic involvement as well as source credibility (Apsler and Sears 1968). For example, Apsler and Sears (1968) found that if message receivers are highly involved in the issue, they are motivated by the warning to defend their position. If, on the other hand, they are not involved in the issue, the warning may have little effect and, in some cases, can even motivate attitude change. The wording of the forewarning- neutral vs. opinionated was also found to influence persuasion outcomes. Infante (1973) found that opinionated forewarning led to the development of more defenses for the topic.

Past studies also identified other approaches to creating belief defenses for a message receiver. A good deal of studies suggested that training a message receiver in refusing skills helps the person to refuse unwanted offers (O'Keefe 2002). While the inoculation-based approach and forewarning approach attempt to create resistance to persuasion by reinforcing the initial attitude, the refusal skills training approach aims at building communication skills in the message receivers. Several studies (Hops et al. 1986; Langlois et al. 1999) found that refusal skills can be created in message receivers by providing training programs. While the discussion of past studies focuses on refusal skill development, this indicates that other skills induced by training could also influence the persuasive communication process.

5.2 Context Factors in Human–Human Communication

The influences of contextual cues on the persuasion process have also been discussed in past studies. Context factors identified in the literature are characteristics of the medium, timing and repetition of the message, and audience reactions (Fig. 5.2).

Research evidence supports that variations in the communication medium affect persuasive outcomes (Hammond 1987; Johnson and Meishcke 1992). Media richness

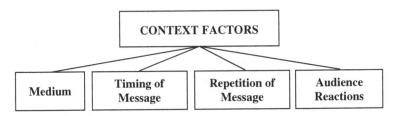

Fig. 5.2   Overview of relevant context factors

theory (Daft and Lengel 1984) proposes that media richness provides important cues that can support the correct processing of information. Media richness, which represents the information carrying capacity of a medium, is a function of (1) the medium's feedback capability; (2) the range of cues available; (3) language variety; and, (4) the medium's ability to convey a personal focus. According to the theory, richer media are better suited for processing complex topics while lean media are appropriate for less equivocal tasks. This theory argues that when communicators select appropriate media in their communications, their task performance is improved. While some empirical findings support the theory (e.g. Daft et al. 1987; Trevino et al. 1990; Kahai and Cooper 2003), a good number of other studies failed to support it (e.g. Dennis and Kinney 1998; El-Shinnawy and Markus 1997; Suh 1999). Currently, findings do not provide a clear picture of which channel is more persuasive. There does not seem to be any general advantage associated with a certain medium.

The impacts of timing and repetition of the message on persuasion were also investigated in past studies. Some studies found that messages delivered temporally close to the point of decision or action can result in maximum effects (O'Keefe 2002) since persuasive effects tend to decay over time (Cook and Flay 1978). However, the rate of decay is not the same in all cases. There is some evidence that persuasion achieved under high elaborated conditions is more likely to be enduring than persuasive outcomes obtained under conditions of low elaboration, i.e., peripheral route processing (Haugtvedt and Strathman 1990). In addition to the timing of a message, the message presentation order can also influence persuasion as discussed earlier in Chap. 4.

In terms of the impacts of message repetition, persuasion tends to be enhanced with the repetition of a message. However, the most common pattern of findings has been an initial increase in persuasion with increased repetition, followed by a decrease in persuasion with further repetition (e.g., Cacioppo and Petty 1979). The positive impact of message repetition is most evident when the message is relatively complex (Cacioppo and Petty 1989).

Audience reactions, for example audience agreement (i.e. consistent and enthusiastic applause) or disagreement with the message (i.e. inconsistent and sparse applause), are also found to be influential in the persuasion process. The impacts of audience reactions differ depending on the message receiver's elaboration likelihood. When elaboration likelihood is low, audience applause acts as a persuasion cue, but when elaboration likelihood is high, the effect of audience cues on persuasion is less prominent (Axsom et al. 1987; Cacioppo and Petty 1982).

5.3 Applying Receiver and Context Factors to Technology

In the recommender system context, some of these factors have been tested and additional cues have been identified. For system user characteristics, the effects of users' intelligence, involvement, and familiarity have been discussed in previous studies. Concerning context factors, product characteristics and recommender system providers' credibility have been examined in existing studies.

5.3.1 Recommender System User Factors

Cues related to the message receiver's intelligence have been tested in some studies. Previous studies examined how system users' product knowledge influenced their evaluations of the system. Empirical evidence suggests that users with low product expertise are more likely to evaluate recommender systems favorably than users with high product expertise. Van Slyke and Collins (1996) found that system users with little product knowledge preferred to use recommender systems in their decision making more than those with high product expertise. The experiment by Urban et al. (1999) also found that participants who did not know much about the products showed a greater preference for the website equipped with a recommender system. Similarly, knowledgeable users were found to be less satisfied with the recommender systems and therefore less reliant on them for their decision making (Spickermann 2001). Further, user expertise is found to be negatively related to perceived ease of use and perceived usefulness of a system (Kamis and Davern 2004). Similar relationships were also found in a recent study. Doong and Wang (2011) found that more product-involved users evaluated the recommendations as less useful for their online shopping.

Furthermore, the system users' product knowledge was found to influence their preference for recommender system types. System users with low product expertise prefer needs-based recommender systems while experts more likely trust feature-based recommender systems (Pereira 2000). Felix and colleagues (2001) also confirmed that users with low product knowledge more likely trust needs-based recommender systems. A recent empirical study (Knijnenburg et al. 2011) further found that novices prefer interaction methods that do not require intimate knowledge of attributes. They derive greater benefits from a non-personalized recommender that just displays the most popular items. In addition, novice decision-makers were found to be more confident in their online decision making when a recommender system is available (O'Hare et al. 2009).

A number of studies have found impacts of user involvement in recommender system evaluations. It needs to be noted that in Elaboration Likelihood Model research, personal relevancy has often been labeled as the receiver's "involvement" with the message topic. But in recommender system context, the term "involvement" has often been used to imply a user's explicit participation in the preference elicitation process. Although "personal relevancy" and message receiver's "involvement" were often used interchangeably in past persuasion studies, the different meanings should be acknowledged in recommender system studies. Zanker et al. (2006) argued that user involvement is an important factor that leads to persuasive outcomes and noted that involved customers more likely relate to recommender systems. Drenner et al. (2008) analyzed data of over 5,000 users and found that the system's prediction error decreased as users rated larger numbers of products; in other words, the system generated more correct user preferences when users were more involved in the preference elicitation process. However, Liang et al. (2006) found that explicit user involvement in the personalization process influences a user's perception of customization, but not his or her overall satisfaction.

Users' familiarity with recommender systems was also found to be an important moderating factor in user and recommender system interactions. Sinha and Swearingen (2002) suggested that users' prior experience with a recommender system plays a role when they decide whether to trust a recommendation from recommender systems. Similarly, users' familiarity with recommender systems was found to increase the intention to adopt a system by enhancing trust in a recommender system's perceived benevolence and integrity but not its competency (Komiak and Benbasat 2006). Recently, Ricci and Nguyen (2007) observed that users who are familiar with traditional web-based recommender systems more likely accept mobile recommender systems.

In addition, some studies found impacts of users' demographic characteristics. For example, a recent study indicates possible gender differences in recommendation acceptance decisions. Doong and Wang (2011) found that women evaluate the perceived usefulness of recommendations to a greater extent than men when they decide whether to accept it or not. Possible effects of users' cultural background were also suggested (Chen and Pu 2008). For instance, a case study of an Austrian spa resort showed that Italian speaking web visitors were twice as likely to use an interactive advisor application than German speaking visitors (Zanker et al. 2008).

5.3.2 Context Factors in Recommender Systems

Product-related characteristics were found to be influential when system users evaluate and interact with recommender systems. Xiao and Benbasat (2007) suggested that product type and complexity moderate the effects of recommender systems on users' decision-making process and outcomes. Some existing findings indicate that online users are more likely to use recommender systems when they shop for experience products. It has been found that recommendations for experience products (wines) were more influential than recommendations for search products (calculators) (Sénécal and Nantel 2003, 2004). This is the case because online users often search for more information (Spiekermann 2001) and are more likely to follow the recommendations from other consumers or organizations for search products (Olshavsky 1985). However, a recent experiment (Ochi et al. 2010) found that there was no significant effect of product type (experience vs. search products) on users' perceived liking, positive feelings towards and perceptions of intelligence of recommender systems.

Some past studies examined the impacts of the recommender system provider (website)'s reputation on users' trust in the recommender system. Xiao and Benbasat (2007) noted that the type and the reputation of the recommender system provider can influence users' trust in system competence, benevolence and integrity. They explained that users may transfer their trust in recommender system providers to the system. Some studies also argued that a website is the first stage during which users develop their trust in recommender systems (Urban et al. 1999) and that website characteristics are important for building trust in recommender systems (West et al. 1999). However, an empirical study (Sénécal and Nantel 2004) that tested the impacts of website type did not support this argument. It found that there was no relationship between the type of website providing the recommender system and users' propensity to follow product recommendations.

Chapter 6
Discussion

Swearingen and Sinha (2001) noted that the ultimate effectiveness of a recommender system depends on factors that go beyond the quality of the algorithm. Nevertheless, recommender system features are oftentimes implemented because they can be implemented. They might be tested in the course of overall system evaluations or usability studies but are rarely assessed in terms of their persuasiveness. Häubl and Murray (2003) demonstrated that recommender systems can indeed have profound impacts on consumer preferences and choice beyond the immediate recommendation. Thus, conceptualizing recommender systems not only as technical artifacts but also as persuasive actors is crucial in understanding their potential impacts.

This book provided a review of the traditional persuasion literature as well as the existing relevant studies in the recommender system realm to provide a conceptual framework for a persuasive recommender system. The review suggests a wide array of recommender system characteristics which could be influential when the system interacts with its users. While the basic framework for persuasive recommender systems was presented in Chap. 1, a comprehensive conceptual model is depicted here based on the review of existing recommender system literature (Fig. 6.1).

The key constructs in the model are the recommender system (source), the recommendation (message), and the system user (receiver) whose interactions are influenced by contextual factors. The persuasive outcomes (effect) are also conceptualized in the model. As presented in the model, current recommender system literature has investigated a number of factors influencing the persuasiveness of a recommender system. For example, the impacts of embodied agents have been examined and various source cues have been implemented. In addition, a number of cues identified in the traditional persuasion literature like message format, message receiver involvement, knowledge as well as familiarity have also been examined in recommender system research. Those examined factors are presented in this conceptual model to provide a current picture of persuasive recommender systems. Therefore, this model gives an idea of the current status of persuasive recommender system research and outlines the key constructs and their

K.-H. Yoo et al., *Persuasive Recommender Systems*, SpringerBriefs in Electrical and Computer Engineering, DOI: 10.1007/978-1-4614-4702-3_6, © The Author(s) 2013

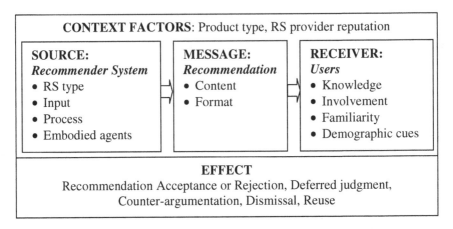

Fig. 6.1 Conceptual model for persuasive recommender systems

relationships. However, it should be used as a starting point only as numerous persuasive factors identified in the human and human communication context (as presented in earlier chapters) have not been applied and tested in a recommender system context. As research continues to investigate and test the factors that can make a recommender system more persuasive, this conceptual model should be updated accordingly. The research gaps identified from the review and the suggestions for future research are further discussed in Chap. 8.

Following the paradigm of "Computers as Social Actors" (Reeves and Nass 1996; Fogg 2003), recent recommender system studies have started emphasizing the social aspects of recommender systems and stress the importance of integrating social cues to create more credible and persuasive systems (Qiu 2006; Wang and Benbasat 2005; Al-Natour et al. 2006). This recognition of recommender systems as social actors has important theoretical implications. Conceptualizing human-recommender system interactions as social exchanges means that important characteristics identified as influential in traditional advice seeking relationships can also be seen as potentially influential in human-recommender system interactions.

Practical implications of this re-conceptualization of recommender systems as social actors are outlined in the following chapter.

Chapter 7
Implications for Recommender System Design

From the marketing point of view, creating credible and persuasive recommender systems is important since recommender systems play similar roles as human salespersons in physical stores who interact with consumers and advise consumers in terms of what to buy (Komiak and Benbasat 2004; Komiak et al. 2005). Thus, creating more sociable and credible recommender systems will help marketers to enhance their e-services.

Although a multitude of persuasive characteristics of recommender systems have been researched until now, a coherent theory of how to design and implement persuasive recommendation applications is still out of sight. For instance, the current state of recommender systems evaluation practice has been examined by Jannach et al. (2010). A small survey on recommender systems research published in the ACM Transactions on Information Systems reveals that 75% of all papers evaluated their contribution by measuring accuracy results on datasets consisting of logs from historic user interactions. A larger quantitative survey that included 330 publications from publication outlets of the Information Systems and the Computer Science communities during the past 5 years confirmed this initial finding. Jannach et al. (2012) classified evaluation practices and measures into the following categories:

- Information Retrieval perspective (IR)
- Machine Learning perspective (ML)
- Decision Support perspective (DS)

The pre-dominant IR practice towards evaluating systems is to measure if the retrieved items are relevant to the user's information need (Manning et al. 2008). Thus, when ground truth of item relevance for specific information needs is known (e.g. determined by votes from an expert panel or derived from past observations) the system's performance can be quantified by the popular Precision, Recall, F1 and rank measures.

Machine Learning focuses on learning models from given example data that most accurately conform to unseen or withheld data. The datasets typically contain past user ratings that are represented on an ordinal or interval scale. For instance,

K.-H. Yoo et al., *Persuasive Recommender Systems*, SpringerBriefs in Electrical and Computer Engineering, DOI: 10.1007/978-1-4614-4702-3_7, © The Author(s) 2013

in parametric models such as linear regression and matrix factorization (Koren et al. 2009) the learner determines those parameter values that minimize an error term representing the deviations between actual and predicted rating values. Error measures such as RMSE or MAE are therefore most commonly used for quantifying the accuracy of a learned model. According to the survey of Jannach et al. (2012) the lion share of all research contributions are either evaluated according to the IR or the ML perspective. The accuracy of recommendations is definitely an important aspect for persuasive applications, notably in the context of message factors. For instance, proposing items that are familiar to the user or propositions that are only slightly discrepant from a user's position have been found to be more persuasive recommendation strategies (compare to Sect. 4.2.1). Therefore, an accurate recommender system is definitely more persuasive than an inaccurate one; however, traditional IR and ML accuracy measures cannot fully capture these aspects of familiarity or slight discrepancy of recommendation content.

Seen from the DS perspective recommender systems are tools that support users in their decision making process. Therefore, quality aspects of a system's decision support capabilities move to the foreground. For instance, the construct of a system's Perceived Usefulness is well known from technology acceptance research (Davis 1989). Online conversion rates are a proxy for measuring commercial success, however, medium and long-term customer satisfaction and the users' propensity to return and recommend the service are more reliable measures that are only rarely applied in recommender system research. Thus, persuasiveness is largely not a guiding objective for developed research prototypes and practical applications. Nevertheless methodological work such as Pu et al. (2011) focusing on a wider array of evaluation objectives are promising signs. Next, we will discuss implications according to the structure of our basic framework for persuasive recommender systems.

7.1 Implications from the Source Perspective

Understanding the influence of source characteristics when evaluating recommender systems has many implications of theoretical and practical importance. From a theoretical perspective, the classic interpersonal communication theories need to be expanded in scope and applied to understand human-recommender system relationships. By applying classic theories, researchers can test and examine various aspects of human-recommender system interactions. Further, while some recommender system-related research exists with respect to source characteristics, these efforts are currently not very systematic and sometimes inconclusive. Clearly, more research is needed in this area such that a strong theoretical framework can be built.

From the practical perspective, understanding recommender systems as social actors whose characteristics influence user perceptions helps system developers

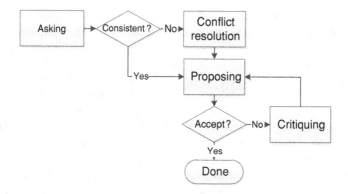

Fig. 7.1 Reference model for conversational recommendation

and designers to better understand user interactions with systems. The way in which preferences are elicited, the way recommendations are derived, and the more insight users have in these processes, the greater perceptions of credibility and the greater the likelihood for a recommendation to be accepted (Gretzel and Fesenmaier 2007). While many practical recommender systems follow a model of one-shot interactions, conversational recommender systems are a branch of systems that support a multi-step communication process. ExpertClerk (Shimazu 2002) is one of these early conversational recommender systems that interact with users by asking and proposing. Figure 7.1 sketches a reference model for conversational recommendation strategies.

Asking denotes the process of eliciting explicit customer requirements as user feedback that can be exploited for various forms of computing recommendations (Zanker et al. 2008). For instance, case-based reasoning strategies determine similarities with past user interactions (i.e. cases) in order to take comparable actions (Shimazu 2002); constraint-based computation relies on encoded sales expertise in order to match abstract customer requirements with technical product features (Zanker et al. 2010).

An example for the latter strategy is the domain independent Advisor framework that has been fielded in different e-commerce and financial services domains (Felfernig et al. 2006). In case user requirements are not satisfiable, i.e. they contradict each other or based on the given product catalog no recommendation is possible, different techniques for conflict resolution exist (Zanker et al. 2010). For instance, explicitly informing users about conflicts and asking them to revise their requirements or providing them with valid alternative sets of requirements are possible strategies. This way the specifics of the search process are communicated to users (Mohr and Bitner 1995; Bechwati and Xia 2003; Sutcliffe et al. 2000) to demonstrate the system's efforts as this will influence credibility perceptions. Otherwise, the system can resolve conflicts on its own, for instance, by proposing those items that fulfill maximal subsets of users' requirements (Jannach 2006; Zanker et al. 2010). Dynamic query interfaces that merge requirements elicitation with propositions within a single interface may be one way to help users feel that they have control over the system as suggested by Schafer (2005). *Critiquing* is a more fine-grained way of learning preference models from users once the system has made some initial propositions. Critiquing interfaces offer users opportunities to formulate detailed negative feedback on proposed items. For instance, the user could reply to an accommodation offer by unit critiques such as "larger rooms", "closer to the center of town" or "lower price". In turn the system then iteratively proposes other recommendations that best possibly consider the received feedback. Burke et al. (1997) were one of the first to propose this way of assisted personalized browsing for exploring the solution space. Since then a variety of extensions such as compound and dynamic critiques have been proposed. See, for instance, Jannach et al. (2010) for a literature survey on these techniques. Compound critiques combine several unit critiques to ensure faster progress in the search space and dynamic critiquing proposes the mining of compound critiques that can be fulfilled by available product alternatives (e.g. "higher service standards but pricier").

In general, interaction strategies that give users control over the process (such as explicit requirements elicitation or critiquing) seem to be highly effective strategies (Xiao and Benbasat 2007; Schafer et al. 2002, 2004; Burke 2002; West et al. 1999; McNee et al. 2003; Konstan and Riedl 2003; Pereira 2000). When generating recommendations, more familiar recommendations with detailed product descriptions (Shinha and Swearingen 2001; Cooke et al. 2002) and explanations regarding the underlying logic of how the recommendation was generated (Wang and Benbasat 2004; Herlocker et al. 2000) would increase users' perceived credibility of the system. Friedrich and Zanker (2011) present a taxonomy that categorizes researched explanation approaches in recommender systems. Basically, explanations are additional information about the recommendations with the purpose of pursuing potential objectives such as transparency or trustworthiness that have been enumerated by Tintarev and Masthoff (2011). Given the complex nature of some statistical techniques to derive recommendations it is however impossible to always explain the underlying logic to users. Therefore, besides white-box explanations describing how the system derived its recommendations, black-box explanation approaches compute justifications arguing why proposed items are plausible recommendations. For instance, these explanations are derived by exploiting the triadic relationship between users, items and tags (Vig et al. 2009; Gedikli et al. 2011) or by employing knowledge bases with domain expertise (Zanker and Ninaus 2010). Recommendation lists, explanations, text descriptions of recommended products along with possibly interactive multimedia applications can all contribute to create virtual product experiences. Jiang and Benbasat (2005) noted that a virtual product experience enhances consumers' product understanding, brand attitude, purchase intention as well as decreases the perceived risks. Adding virtual experiences of products enables the users not only to have a better understanding of the recommended products but also to inspire greater attention, interest and enjoyment.

While the aforementioned aspects of source characteristics addressed the interaction strategy of recommender systems, many empirical findings about persuasive traits of recommender systems focus on the visual design of the system.

The challenge lies in finding ways to translate source characteristics such as similarity, likeability and authority into concrete design features that fit within the context of recommender systems. For instance, presenting third party seals signaling the authority of the system can increase the overall credibility of systems. Similarity between recommender systems and users can be implemented by the use of needs-based questions that elicit users' product preferences and the decision strategies (Xiao and Benbasat 2007). Manipulating personalities (e.g. extraversion or introversion) of recommender systems to match with those of users by varying communication style and voice characteristics was also suggested by Hess et al. (2005) and Moon (2002). One way in which some characteristics can be more easily implemented is by adding an embodied agent to the system interface.

The options for personifications range from artificial cartoon-like or anthropomorphic characters to depictions of real humans. Furthermore, the degree of animation ranges from simple picture files to full-fledged three dimensional (3D) characters being capable of mimics and gestures. The embodied agent serves as the representative of the system and, thus, emphasizes the social role of the system as the advice giver (Yoo and Gretzel 2009). Voice interfaces can be another way to translate source characteristics into credibility-evoking recommender system design. In principle, the technical basis for different forms of multimodal interaction between users and online applications via voice, free text and on-the fly generated visual interfaces is available as already noted by Lucente (2000). Comparable to the early ELIZA system (Weizenbaum 1966) virtual natural language shop assistants—termed lingubots—parse user input for keywords and patterns in order to apply transformation rules to rephrase the user input and to retrieve appropriate textual phrases to reply. Depending on the engineering effort to develop rule sets with canned text the lingubot will mimic more or less "intelligent" behavior. The following shortened dialogue taken from Zanker et al. (2009) serves as an example that disguises still limited language capabilities of lingubots:

> Lingubot: *My name is Frank. How can I assist you?*
> User: *Do you sell products?*
> Lingubot: *I don't think I was designed for the purpose of making sales.*
> User: *Do you sell wine?*
> Lingubot: *Robots don't drink any liquids, not even wine.*

While the lingubot of a major technology provider correctly replies to the first question, the more specific second question shows the limited depth of the lingubot's rule base. The reply simply matches the keyword "wine" and ignores the verb "sell". Thus, it becomes counter effective if high expectations towards natural language interactions are created that cannot be met by system functionality as has already been discussed in Sect. 3.2.2.4. In case effort-intense natural language interfaces are out of budget, forms based interactive dialogues with personified characters can serve as a cost-efficient compromise (Felfernig et al. 2006; Zanker et al. 2009).

Summarizing, the interaction between recommender system and user has highest practical relevance for developing persuasive recommendation applications. One very promising piece of work into that direction is Mahmood et al. (2010). They deconstruct the interaction between system and user into a state model and propose to learn optimal interaction strategies. Two types of states exist: view states (e.g. input forms for making a request and information pages), where users can provide some input or make a request, and system decision points. In the latter states the system has to make choices between alternative conversational moves, for instance, proposing an item or asking the user to provide more detailed requirements. Mahmood et al. apply Markovian learning models to determine the optimal policy in order to reach goal states such as user conversion when the history of past interactions is known.

7.2 Implications from the Message Perspective

Recommender system designers should also pay attention to the display format of the recommendations (Swearingen and Sinha 2001; Yoon and Lee 2004). Navigational efficacy and design familiarity and attractiveness need to be considered when the recommendations are presented to users. In addition, various decision biases influence users' appreciation of recommended items and challenge the assumption that users take rational choices and decisions when shopping online. Simon (1959) ascribed these limitations of humans' cognitive processing capabilities with the term "bounded rationality", i.e. human decision makers do not maximize objective utility functions. Effects that can partly explain these phenomena of bounded rationality in online choice tasks are for instance position effects, decoy effects and framing effects.

According to Abed (1991) primacy effects in online choice and reading tasks, i.e. the first items in a list catch the most attention and are therefore best remembered, are explained by the culturally dependent reading direction from left to right and from top to bottom in western countries. For instance, empirical findings from Granka et al. (2004) report a strong primacy effect for clicking behavior in online search. Comparably, we can assume that in recommendation lists the topmost (vertical ordering) and the leftmost (horizontal ordering) items do receive more user attention and will therefore be biased towards higher click-through and conversion rates. Thus, when evaluating the popularity of items the positions of their presentations also need to be considered.

Decoy effects (Teppan and Felfernig 2009) occur when so-termed decoy items are present in choice sets. They increase the attraction of a target item and/or decrease the attraction of a competitor item. Figure 7.2 sketches a very simple example. Two items, A and B, differ by opposing attribute evaluations on two arbitrary characterizing dimensions (attribute 1 and 2). Item A scores better than B

Fig. 7.2 Example of decoy
products

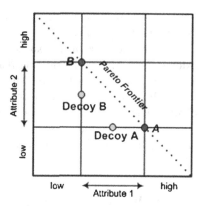

with respect to attribute 1 but does worse with respect to attribute 2. Thus, both items are on the same Pareto Frontier. This means that by switching from one frontier item to another, one cannot improve an attribute dimension without accepting to deteriorate another dimension. The item Decoy A increases the attraction of item A, because A is always a better choice than Decoy A (A and Decoy A possess the same utility w.r.t. attribute 2, but A is clearly better on attribute 1). However, when comparing Decoy A with item B there is no clear winner (Decoy A is better on attribute 1, but worse on attribute 2). Consequently, in choice sets where users have to decide between items A, B and Decoy A, item A will experience a positive bias due to the presence of Decoy A.

The framing effect describes the notion of people deciding differently on the same set of options based on the decision frame in which options are presented (Tversky and Kahnemann 1986). Kahneman and Tverski (1979) explain the framing effect by the concept of loss aversion which is also the basis for their Prospect Theory. In summary, people tend to experience (possible) losses more than (possible) gains. This triggers people to react risk averse when options are described in terms of gains and risk seeking when the same options are described in terms of losses. In recommendation scenarios the situational context of users is typically a buying situation that is framed as a decision about different gains. Therefore, we can assume that users will predominantly act to be risk-averse and choose the less risky option even if expected utilities are lower.

Teppan and Felfernig (2010) performed a variety of different experimental studies where they explored position and decoy effects on result pages of recommender systems for different product domains. Notably, their proposition of a decoy filter is a very valuable practical contribution. Their algorithm for decoy minimization (DM) identifies misleading decoy effects by comparing objective utility values with context-dependent utilities (i.e. based on the set of items that are recommended together). Once the algorithm determines decoy items in recommendation sets, it can either simply remove them or neutralize them by adding additional decoy items. In the latter case items whose choice probabilities are lowered due to the existence of a decoy item are strengthened by additionally introduced decoy items that support them.

Very recent work of Teppan and Zanker (2012) builds on these studies and extends them by also considering risk aversion strategies of users when appreciating items as well as by observing interactions and relative strengths of these effects. It actually turned out that risk aversion strategies biased study participants' choices much stronger than decoy and position effects. Thus, in order to ensure objective decision making and elicitation of true user preferences, understanding and control of the aforementioned decision biases is crucial.

7.3 Implications from the Receiver Perspective

Empirical findings support the hypothesis that receiver characteristics such as involvement and domain expertise are relevant receiver characteristics that influence the perception of persuasive system traits as discussed in Chap. 5. Therefore, these user factors need to be observed and considered when designing the interaction experience. Conversational recommender systems need to offer different modes of interaction. For instance, the DIETORECS project (Fesenmaier et al. 2003) from the tourism domain is one of the earliest examples for supporting different modes of preference elicitation and decision making. Users can, for instance, select a destination based on inspiring pictures or specify concrete needs and restrictions. In technical product domains such as consumer electronics different levels of abstraction of product features need to be applied in the conversation with users. For instance, inexperienced notebook shoppers might not know the meaning of abbreviations such as HDMI, but should be asked if they want to connect external devices for playing high-definition multimedia content. In contrast, domain experts might find concise search interfaces for specifying these technical product features most efficient and useful.

Furthermore, in the context of Web 2.0 applications and the availability of information about users' social networks additional opportunities for developing persuasive recommender systems emerge. While collaborative recommendation algorithms have been extended to exploit social and trust networks for computing user neighborhoods and deriving recommendations, the persuasive potential of Web 2.0 has not yet been fully explored in recommender systems research. For instance, recommender systems could model the dynamics in social groups and apply game theoretic considerations about whom to recommend what and when. However, the discussion about furthering current research in the context of persuasive recommendation and also related ethical considerations will be continued in the next chapter.

Chapter 8
Directions for Future Research

While existing studies have identified and tested a number of influential characteristics in human-recommender system advice seeking relationships, many potential characteristics suggested by general communication theories such as authority, caring, and humor have not been examined. Those unexamined characteristics need to be successfully implemented and also empirically tested in future recommender system studies.

The identified and tested characteristics also need to be more precisely examined. The effects of source characteristics on judgments of source credibility are often found to be complex rather than linear in previous studies conducted in human-human advice seeking contexts (O'Keefe 2002). Since situational factors, individual differences and product type can also play a significant role in determining the recommender system's credibility, relationships will have to be specifically tested for specific recommender systems to provide accurate input for design considerations.

In addition, there can be additional persuasive characteristics that might not be prominent in influencing advice seeking relationships among human actors but are important aspects to be considered in the realm of recommender systems. For instance, anthropomorphism of the technology has been identified as an important characteristic that influences interactions with technologies (Koda 1996; Nowak and Biocca 2003; Yoo 2010) while it is of course not a critical characteristic in interactions among human actors. The realness of interface agents can also be considered as a potentially influential source cue. There is some evidence that users are less likely to respond socially to a poor implementation of a human-like software character than to a good implementation of a dog-like character (Kiesler et al. 1996). In future research, such additional source cues need to be identified and tested.

Some of the influential characteristics have been tested in isolation from another. In order to investigate interaction effects, different cues should be tested simultaneously if it is possible to implement them at the same time and also to examine the relationship among source, message as well as receiver factors. This will help with understanding the relationships among various factors.

K.-H. Yoo et al., *Persuasive Recommender Systems*, SpringerBriefs in Electrical and Computer Engineering, DOI: 10.1007/978-1-4614-4702-3_8, © The Author(s) 2013

Overall, the literature presented in this book suggests that there is a great need for research in this area. It also suggests that new methodologies might have to be developed to investigate influences that happen at a sub-conscious level. Especially a greater emphasis on behavioral measures of recommendation acceptance seems to be warranted if the persuasiveness of recommender systems is to be evaluated. There is also a critical lack of qualitative research in the recommender system realm that could help build theory based on rich under-standings of the user-system interaction. Further, recommender systems are nowadays often accessed through mobile devices. This changes the interaction process considerably. While there have been studies on mobile recommender systems (Ricci 2011), comprehensive evaluations of interactions and persuasion in mobile contexts are currently missing from the literature. As Ricci (2011) noted, usability is influenced by particular characteristics of the mobile devices. For instance, the input and interaction capabilities are often limited on mobile devices (Ricci 2011). Small size of display was found to lessen the effectiveness of users' task completion (Jones and Marsden 2005). In addition, tasks completed on mobile devices can be frequently interrupted since mobile users access information on the move. Considering these particular aspects affecting mobile interactions with recommender systems, persuasive cues and their effects in mobile contexts should be examined in future research. Another aspect to consider is that technology use is often social (Gretzel 2011). This means that future research should also consider situations with multiple receivers and the dynamics that emerge from their interactions among themselves and with the recommender system.

Recommender systems are here to stay and will only increase in importance with an ever greater amount of information available online and an increasing need for personalized solutions. Research that can help improve the interactions recommender systems facilitate is essential for driving recommender system developments forward. One of the challenges with increasing their persuasiveness is however, the question of which persuasion attempts are ethical and which are not. Users often see recommender systems as objective in their choices of products to recommend. This is of course not always the case. While user preferences play a fundamental role, marketer or provider preferences are usually also incorporated into such systems. Recommender system developers and researchers will have to decide where they draw the line between persuasion and manipulation. Ultimately, users will evaluate their satisfaction with the choices they made based on the recommendation. Therefore, manipulation will not pay off in the long run.

References

Abed, F. (1991). Cultural influences on visual scanning patterns. *Journal of Cross-Cultural Psychology, 22*, 525–534.

Addington, D. W. (1971). The effect of vocal variations on ratings of source credibility. *Speech Monographs, 38*, 242–247.

Aksoy, L., Bloom, P. N., Lurie, N. H., & Cooil, B. (2006). Should recommendation agents think like people? *Journal of Service Research, 8*(4), 297–315.

Al-Natour, S., Benbasat, I., & Cenfetelli, R. T. (2006). The role of design characteristics in shaping perceptions of similarity: The case of online shopping assistants. *Journal of Association for Information Systems, 7*(12), 821–861.

Andersen, K. E., & Clevenger, T., Jr. (1963). A summary of experimental research in ethos. *Speech Monographs, 30*, 59–78.

Ansari, A., Essegaier, S., & Kohli, R. (2000). Internet recommendation systems. *Journal of Marketing Research, 37*(3), 363–375.

Apsler, R., & Sears, D. O. (1968). Warning, personal involvement, and attitude change. *Journal of Personality and Social Psychology, 9*, 162–166.

Aronson, E., Turner, J. A., & Carlsmith, J. M. (1963). Communicator credibility and communication discrepancy as determinants of opinion change. *Journal of Abnormal and Social Psychology, 67*, 31–36.

Atkin, C., & Block, M. (1983). Effectiveness of celebrity endorsers. *Journal of Advertising Research, 23*(1), 57–61.

Atkinson, D. R., Winzelberg, A., & Holland, A. (1985). Ethnicity, locus of control for family planning, and pregnancy counselor credibility. *Journal of Counseling Psychology, 32*, 417–421.

Axsom, D., Yates, S. M., & Chaiken, S. (1987). Audience response as a heuristic cue in persuasion. *Journal of Personality and Social Psychology, 53*, 30–40.

Baker, M. J., & Churchill, G. A., Jr. (1977). The impact of physically attractive models on advertising evaluations. *Journal of Marketing Research, 14*(4), 538–555.

Bart, Y., Shankar, V., Sultan, F., & Urban, G. L. (2005). Are the drivers and role of online trust the same for all web sites and consumers? A large scale exploratory and empirical study. *Journal of Marketing, 69*(4), 133–152.

Basartan, Y. (2001). Amazon Versus the Shopbot: An Experiment About How to Improve the Shopbots. Ph.D. Summer Paper, Carnegie Mellon University, Pittsburgh, PA.

Bauernfeind, U., & Zins, A. H. (2006). The perception of exploratory browsing and trust with recommender websites. *Information Technology and Tourism, 8*(2), 121–136.

Bechwati, N. N., & Xia, L. (2003). Do computers sweat? The impact of perceived effort of online decision aids on consumers' satisfaction with the decision process. *Journal of Consumer Psychology, 13*(1–2), 139–148.

K.-H. Yoo et al., *Persuasive Recommender Systems*, SpringerBriefs in Electrical and Computer Engineering, DOI: 10.1007/978-1-4614-4702-3, © The Author(s) 2013

Becker, B. J. (1986). Influence again: An examination of reviews and studies of gender differences in social influence. In J. S. Hyde & M. C. Linn (Eds.), *The psychology of gender: Advances through meta-analysis* (pp. 178–209). Baltimore: Johns Hopkins University Press.

Belch, G., & Belch, M. (2009). *Advertising and promotion: An integrated marketing communications perspective.* McGraw-Hill.

Bharti, P., & Chaudhury, A. (2004). An empirical investigation of decision-making satisfaction in web-based decision support systems. *Decision Support Systems, 37*(2), 187–197.

Bickman, L. (1974). The social power of a uniform. *Journal of Applied Social Psychology, 4,* 47–61.

Bonhard, P., & Sasse, M. A. (2005). I thought it was terrible and everyone else loved it—A new perspective for effective recommender system design. In *Proceedings of the 19th British HCI Group Annual Conference*, Napier University (pp. 251–261). Edinburgh, UK 5–9 Sept 2005.

Buller, D. B., & Burgoon, J. K. (1996). Interpersonal deception theory. *Communication Theory, 6,* 203–242.

Burgoon, J. K., Birk, T., & Pfau, M. (1990). Nonverbal behaviors, persuasion, and credibility. *Human Communication Research, 17,* 140–169.

Burgoon, J. K., Dunbar, N. E., & Segring, C. (2002). Nonverbal influence. In J. P. Dillard & M. Pfau (Eds.), *Persuasion handbook: Developments in theory and practice* (pp. 445–473). Thousand Oaks: Sage Publications.

Burke, R. (2002). Hybrid recommender systems: Survey and experiments. *User Modeling and User-Adapted Interaction, 12*(4), 331–370.

Burke, R., Hammond, K., & Young, B. (1997). The find me approach to assisted browsing. *IEEE Expert, 4*(12), 32–40.

Bryant, J., Brown, D., Silberberg, A. R., & Elliott, S. M. (1981). Effects of humorous illustrations in college textbooks. *Human Communication Research, 8,* 43–57.

Byrne, D. (1971). *The attraction paradigm.* New York: Academic.

Byrne, D., & Rhamey, R. (1965). Magnitude of positive and negative reinforcements as a determinant of attraction. *Journal of Personality and Social Psychology, 2,* 884–889.

Cacioppo, J. T., & Petty, R. E. (1979). Effects of message repetition and position on cognitive responses, recall, and persuasion. *Journal of Personality and Social Psychology, 37,* 97–109.

Cacioppo, J. T., & Petty, R. E. (1982). The need for cognition. *Journal of Personality and Social Psychology, 42,* 116–131.

Cacioppo, J. T., & Petty, R. E. (1989). Effects of message repetition on argument processing, recall, and persuasion. *Basic and Applied Social Psychology, 10,* 3–12.

Carli, L. L., Ganley, R., & Pierce-Otay, A. (1991). Similarity and satisfaction in roommate relationships. *Personality and Social Psychology Bulletin, 17*(4), 419–426.

Chance, P. (1975). Ads without answers make brain itch. *Psychology Today, 9,* 78.

Chang, K.-J., & Gruner, C. R. (1981). Audience reaction to self-disparaging humor. *Southern Speech Communication Journal, 46,* 419–426.

Chen, L., & Pu, P. (2008). A cross-cultural user evaluation of product recommender interfaces. In *Proceedings of the 2008 ACM Conference on Recommender Systems* (RecSYS'08), Lausanne, Switzerland.

ChoiceStream (2009). 2009 ChoiceStream personalization survey. http://www.choicestream.com/surveyresults/. Accessed 14 Feb 2010.

Cialdini, R. B. (1993). *Influence: The psychology of persuasion.* New York: William Morrow.

Cialdini, R. B. (1994). Interpersonal influence. In S. Shavitt & T. C. Brock (Eds.), *Persuasion: Psychological insights and perspective* (pp. 195–217). Massachusetts, Allyn and Bacon: Needhan Heights.

Cialdini, R. B., Vincent, J. E., Lewis, S. K., Catalan, J., Wheeler, D., & Darby, B. L. (1975). Reciprocal concessions procedure for inducing compliance: The door-in-the-face technique. *Journal of Personality and Social Psychology, 31,* 206–215.

Cook, T. D., & Flay, B. R. (1978). The persistence of experimentally induced attitude change. In Berkowitz (Ed.), *Advances in experimental social psychology* (Vol. 11, pp. 1–57). New York: Academic.

Cooke, A. D. J., Sujan, H., Sujan, M., & Weitz, B. A. (2002). Marketing the unfamiliar: the role of context and item-specific information in electronic agent recommendations. *Journal of Marketing Research, 39*(4), 488–497.

Cooper, J., Bennett, E. A., & Sukel, H. L. (1996). Complex scientific testimony: How do jurors make decisions? *Law and Human Behavior, 20,* 379–394.

Cosley, D., Lam, S. K., Albert, I., Konstan, J., & Riedl, J. (2003). Is seeing believing? How recommender systems influence users' opinions. In *Proceedings of CHI 2003: Human Factorsin Computing Systems* (pp. 585–592). New York: ACM Press.

Cowell, A. J., & Stanney, K. M. (2005). Manipulation of non-verbal interaction style and demographic embodiment to increase anthropomorphic computer character credibility. *International Journal of Human-Computer Studies, 62,* 281–306.

Daft, R. L., & Lengel, R. H. (1984). Information richness: a new approach to managerial behavior and organizational design. In L. L. Cummings and B. M. Staw (Eds.), *Research in organizational behavior* (pp. 191–233). Homewood, IL: JAI Press.

Daft, R. L., Lengel, R. H., & Trevino, L. K. (1987). Message Equivocality, Media Selection and Manager Performance: Implications for Information Systems. MIS Quarterly, September, pp. 355–366.

Davis, F. D. (1989). Perceived usefulness, perceive ease of use, and user acceptance of information technology. *MIS Quarterly, 13*(3), 319–340.

Dean, R. B., Austin, J. A., & Watts, W. A. (1971). Forewarning effects in persuasion: Field and classroom experiments. *Journal of Personality and Social Psychology, 18*(2), 210–221. doi: 10.1037/h0030835.

DeJong, W. (1979). An examination of self-perception mediation of the foot-in-the-door effect. *Journal of Personality and Social Psychology, 37,* 2221–2239.

Delgado-Ballester, E. (2004). Applicability of a brand trust scale across product categories: A multigroup invariance analysis. *European Journal of Marketing, 38*(5/6), 573–592.

Delia, J. G. (1975). Regional dialect, message acceptance, and perceptions of the speaker. *Central States Speech Journal, 26,* 188–194.

Dennis, A. R., & Kinney, S. T. (1998). Testing media richness theory in the new media: the effects of cues, feedback, and task equivocality. *Information Systems Research, 9*(3), 256–274.

Dijkstra, J. J., Liebrand, W. B. G., & Timminga, E. (1998). Persuasiveness of Expert Systems. *Behaviour and Information Technology, 17*(3), 155–163.

Doong, H., & Wang, H. (2011). Do males and females differ in how they perceive and elaborate on agent-based recommendations in Internet-based selling? *Electronic Commerce Research and Applications, 10*(5), 595–604.

Drenner, S., Sen S., & Terveen, L. (2008). Crafting the initial user experience to achieve community goals. In *2008 ACM Conference on Recommender Systems* (RecSys'08), (pp. 187–194). New York: ACM.

Eagly, A. H., Ashmore, R. D., Makhijani, M. G., & Longo, L. C. (1991). What is beautiful is good, but…: A meta-analytic review of research on the physical attractiveness stereotype. *Psychological Bulletin, 110,* 109–128.

Eagly, A. H., & Carli, L. L. (1981). Sex of researchers and sex-typed communications as determinants of sex differences in influenceability: A meta-analysis of social influence studies. *Psychological Bulletin, 90,* 1–20.

Eagly, A. H., & Chaiken, S. (1975). An attribution analysis of the effect of communicator characteristics on opinion change: The case of communicator attractiveness. *Journal of Personality and Social Psychology, 32*(1), 136–144.

Eagly, A. H., Wood, W., & Chaiken, S. (1978). Causal inferences about communicators and their effect on opinion change. *Journal of Personality and Social Psychology, 36*(424), 435.

El-Shinnawy, M., & Markus, M. L. (1997). The poverty of media richness theory: Explaning people's choice of electronic mail vs. voice mail. *International Journal of Human-Computer Studies, 46*, 443–467.

Engstrom, E. (1994). Effects of nonfluencies on speakers' credibility in newscast settings. *Perceptual and Motor Skills, 78*, 739–743.

Evans, R. I., Rozelle, R. M., Lasater, T. M., Dembroski, T. M., & Allen, B. P. (1970). Fear arousal, persuasion, and actual versus implied behavioral change: New perspective utilizing a real-life dental hygiene program. *Journal of Personality and Social Psychology, 16*, 220–248.

Fasolo, B., McClelland, G. H., & Lange, K. A. (2005). The effect of site design and interattribute correlations on interactive web-based decisions. In C. P. Haughvedt, K. Machleit, & R. Yalch (Eds.), *Online consumer psychology: Understanding and influencing behavior in the virtual world* (pp. 325–344). Mahwah: Lawrence Erlbaum Associates.

Felfernig, A., Friedrich, G., Jannach, D., & Zanker, M. (2006). An Integrated environment for the development of knowledge-based recommender applications. *International Journal of Electronic Commerce, 11*(2), 11–34. doi:10.2753/JEC1086-4415110201.

Felix, D., Niederberger, C., Steiger, P., and Stolze, M. (2001). Featur-oriented vs. needs-oriented product access for non-expert online shoppers. In B. Schmid, K. Stanoevska-Slabeva, and V. Tschammer-Zurich (Eds.), *Towards the E-Society: E-Commerce, E-Business, and E-Government* (pp. 399–406). New York: Springer.

Fesenmaier, D. R., Ricci, F., Schaumlechner, E., Wöber, K. W. & Zanella, C. (2003). DIETORECS: Travel advisory for multiple decision styles. In *Proceedings of the International Conference on Information and Communication Technologies in Tourism* (ENTER) (pp. 232–241) Springer, Helsinki, Finland.

Fink, E. L., Kaplowitz, S. A., & Bauer, C. L. (1983). Positional discrepancy, psychological discrepancy, and attitude change: Experimental tests of some mathematical models. *Communication Monographs, 50*, 413–430.

Flanagin, A. J., & Metzger, M. J. (2003). The perceived credibility of personal Web page information as influenced by the sex of the source. *Computers in Human Behavior, 19*(6), 683–701.

Fleshler, H., Ilardo, J., & Demoretcky, J. (1974). The influence of field dependence, speaker credibility set, and message documentation on evaluations of speaker and message credibility. *Southern Speech Communication Journal, 39*, 389–402.

Fogg, B. J. (2003). *Persuasive technology: Using computers to change what we think and do*. San Francisco: Morgan Kaufmann.

Fogg, B. J., Lee, E., & Marshall, J. (2002). Interactive technology and Persuasion. In J. P. Dillard & M. Pfau (Eds.), *Persuasion handbook: Developments in theory and practice* (pp. 765–797). London: Sage.

Fogg, B. J., & Nass, C. (1997). Silicon sycophants: Effects of computers that flatter. *International Journal of Human-Computer Studies, 46*(5), 551–561.

Frantz, J. P. (1994). Effect of location and procedural explicitness on user processing of and compliance with product warnings. *Human Factors, 36*, 532–546.

Freedman, J. L., & Fraser, S. C. (1966). Compliance without pressure: The foot-in-the-door technique. *Journal of Personality and Social Psychology, 4*, 195–202.

Friedman, H., & Friedman, L. (1979). Endorser effectiveness by product type. *Journal of Advertising Research, 19*(5), 63–71.

Gatignon, H., & Robertson, T. S. (1991). *Innovative decision processes*. Englewood Cliffs: Prentice Hall.

Gedikli, F., Ge, M., & Jannach, D. (2011). Understanding recommendations by reading the clouds. *12th International Conference on Electronic Commerce and Web Technologies (EC-Web), Springer*, (pp. 196–208).

Giffen, K., & Ehrlich, L. (1963). Attitudinal effects of a group discussion on a proposed change in company policy. *Speech Monographs, 30*, 377–379.

Giles, H., & Coupland, N. (1991). *Language: Contexts and consequences*. Pacific Grove: Brooks/ Cole.

Gilkinson, H., Paulson, S. F., & Sikkink, D. E. (1954). Effects of order and authority in an argumentative speech. *Quarterly Journal of Speech, 40*, 183–192.

Gilly, M. C., Graham, J. L., Wolfinbarger, M. F., & Yale, L. J. (1998). A dyadic study of personal information search. *Journal of the Academy of Marketing Science, 26*(2), 83–100.

Granka, L. A., Joachims, T., & Gay, G. (2004). Eyetracking analysis of user behavior in WWW search. In *Proceedings of the 27th Annual International ACM SIGIR Conference on Research and Development in Information Retrieval* (pp. 478–479). ACM.

Gretzel, U. (2004). Consumer responses to preference elicitation processes in destination recommendation systems. Doctoral dissertation, University of Illinois at Urbana-Champaign.

Gretzel, U. (2006). Narrative Design for Travel Recommender Systems. In D. R. Fesenmaier, H. Werthner, & K. W. Wöber (Eds.), *Destination recommendation systems: Behavioural foundations and applications* (pp. 190–201). Cambridge: CABI.

Gretzel, U. (2011). Intelligent systems in tourism: A social science perspective. *Annals of Tourism Research, 38*(3), 757–779.

Gretzel, U., & Fesenmaier, D. R. (2007). Persuasion in recommender systems. *International Journal of Electronic Commerce, 11*(2), 81–100.

Gruner, C. R., & Lampton, W. E. (1972). Effects of including humorous material in a persuasive sermon. *Southern Speech Communication Journal, 38*, 188–196.

Gulley, H. E., & Berlo, D. K. (1956). Effect of intercellular and intracellular speech structure on attitude change and learning. *Speech Monographs, 23*, 288–297.

Gundersen, D. F., & Hopper, R. (1976). Relationships between speech delivery and speech effectiveness. *Communication Monographs, 43*, 158–165.

Hammond, S. L. (1987). Health advertising: The credibility of organizational sources. *Communication Yearbook, 10*, 613–628.

Han, S., & Shavitt, S. (1994). Persuasion and culture: Advertising appeals in individualistic and collectivistic societies. *Journal of Experimental Social Psychology, 30*, 326–350.

Harkins, S. G., & Petty, R. E. (1981). The multiple source effect in persuasion: The effects of distraction. *Personality and Social Psychology Bulletin, 4*, 627–635.

Harkins, S. G., & Petty, R. E. (1987). Information utility and the multiple source effect. *Journal of Personality and Social Psychology, 52*, 260–268.

Harmon, R. R., & Coney, K. A. (1982). The persuasive effects of source credibility in buy and lease situations. *Journal of Marketing Research, 19*(2), 255–260.

Häubl and Murray. (2003). Preference construction and persistence in digital marketplaces: The role of electronic recommendation agents. *Journal of Consumer Psychology, 13*(1&2), 75–91.

Haugtvedt, C.P., & Strathman, A. J. (1990). Situational product relevance and attitude persistence. *Advances in Consumer Research, 17*, 766–769.

Haugtvedt, C.P., & Wegener, D. T. (1994). Message order effects in persuasion: An attitude strength perspective. *Journal of Consumer Research, 21*, 205–218.

Herlocker, J., Konstan, J. A., & Riedl, J. (2000). Explaining collaborative filtering recommendations. *Proceedings of the 2000 ACM Conference on Computer Supported Cooperative Work*, Philadelphia, pp. 241–250.

Hess, T. J., Fuller, M. A., & Mathew, J. (2005). Involvement and decision-making performance with a decision aid: The influence of social multimedia, gender, and playfulness. *Journal of Management Information Systems, 22*(3), 15–54.

Hewgill, M. A., & Miller, G. R. (1965). Source credibility and response to fear-arousing communications. *Speech Monographs, 32*, 95–101.

Hoffman, D. L., & Novak, T. P. (1996). Marketing in hypermedia computer-mediated environments: Conceptual foundations. *Journal of Marketing, 60*(July), 50–68.

Hofling, C. K., Brotzman, E., Dalrymple, S., Graves, N., & Pierce, C. M. (1966). An experimental study of nurse-physician relationships. *Journal of Nervous and Mental Disease, 143*, 171–180.

Hogg, M. A., CooperShaw, L., & Holzworth, D. W. (1993). Group prototypically and depersonalized attraction in small interactive groups. *Personality and Social Psychology Bulletin, 19*(4), 452–465.

Holzwarth, M., Janiszewski, C., & Neumann, M. M. (2006). The influence of avatars on online cosumer shopping behavior. *Journal of Marketing, 70*(October), 19–36.

Hops, H., Weissman, W., Biglan, A., Thompson, R., Faller, C., & Severson, H. H. (1986). A taped situation test of cigarette refusal skill among adolescents. *Behavioral Assessment, 8*, 145–154.

Horai, J., Naccari, N., & Fatoullah, E. (1974). The effects of expertise and physical attractiveness upon opinion agreement and liking. *Sociometry, 37*, 601–606.

Hovland, C. I., Janis, I. L., & Kelley, H. H. (1953). *Communication and persuasion.* New Haven: Yale University Press.

Hovland, C. I., & Mandell, W. (1952). An experimental comparison of conlusion drawing by the communicator and by the audience. *Journal of Abnormal and Social Psychology, 47*, 581–588.

Hovland, C. I., & Weiss, W. (1951). The influence of source credibility on communication effectiveness. *The Public Opinion Quarterly, 15*(4), 635–650.

Infante, D. A. (1973). Forewarning in persuasion: Effects of opinionated language and forewarner and speaker authoritativeness. *Western Speech, 37*(3), 185–195.

Jackson, J. M. (1987). Social impact theory: A social forces model of influence. In B. Mullen & G. R. Goethals (Eds.), *Theories of group behavior* (pp. 111–124). New York: Springer Verlag.

Jackson, J. Z., & Devine, P. G. (2000). Attitude importance, forewarning of message content, and resistance to persuasion. *Basic and Applied Social Psychology, 22*, 19–29.

Jannach, D. (2006). Techniques for Fast Query Relaxation in Content-Based Recommender Systems. *29th Annual German Conference on AI (KI), Springer*, (pp. 9–63).

Jannach, D., Zanker, M., Felfernig, A., & Friedrich, G. (2010). *Recommender systems an introduction.* Cambridge: Cambridge University Press.

Jannach, D., Zanker, M., Ge, M. & Groening, M. (2012). Recommender systems in computer science and information systems—A landscape of research. *E-Commerce and Web Technologies (EC-Web), Springer, LNBIP, 123*, 76–87.

Jiang, J. J., Klein, G., & Vedder, R. G. (2000). Persuasive expert systems: The influence of confidence and discrepancy. *Computers in Human Behavior, 16*, 99–109.

Jiang, Z., & Benbasat, I. (2005). Virtual product experience: Effects of visual and functional control of products on perceived diagnosticity and flow in electronic shopping. *Journal of Management Information Systems, 21*(3), 111–148.

Johnson, B. T., & Eagly, A. H. (1989). Effects of involvement on persuasion: A meta-analysis. *Psychological Bulletin, 106*, 290–314.

Jones, M., & Marsden, G. (2005). *Mobile interaction design.* NY: Wiley.

Johnson, J. D., & Meishcke, H. (1992). Differences in evaluations of communication channels for cancer-related information. *Journal of Behavioral Medicine, 15*, 429–445.

Kahai, S. S., & Cooper, R. B. (2003). Exploring the core concepts of media richness theory: the impact of cue multiplicity and feedback immediacy on decision quality. *Journal of Management Information Systems, 20*(1), 263–299.

Kahneman, D., & Tversky, A. (1979). Prospect theory: An analysis of decision under risk. *Econometrica, 47*, 263–291.

Kamis, A., & Davern, M. J. (2004). Personalizing to product category knowledge: Exploring the mediating effect of shopping tools on decision confidence. *Proceedings of the* 37th *Annual Hawaii International Conference on System Sciences*, Big Island, Hi, January 2004.

Kang, M., & Gretzel, U. (2012). Effects of podcast tours on tourist experiences in a national park. *Tourism Management, 33*(2), 440–455.

Karlins, M., & Abelson, H. I. (1970). *How opinions and attitudes are changed* (2nd ed.). New York: Springer Verlag.

Kelman, H. C., & Hovland, C. I. (1953). Reinstatement of the communicator in delayed measurement of opinion change. *Journal of Abnormal Psychology, 48*(3), 327–335.

Kim, D.-Y., & Morosan, C. (2006). Playfulness on website interactions: Why can travel recommendation systems not be fun? In D. R. Fesenmaier, H. Werthner, & K. W. Wöber (Eds.), *Destination recommendation systems: Behavioural foundations and applications* (pp. 190–201). Cambridge: CABI.

Kiesler, S., Sproull, L., & Waters, K. (1996). A prisoner's dilemma experiment on cooperation with people and human-like computers. *Journal of Personality and Social Psychology, 70*(1), 47–65.

Knijnenburg, B. P., Reijmer, N. J. M., & Willemsen, M. C. (2011). Each to his own: How different users call for different interaction methods in recommender systems. In *Proceedings of the Fifth ACM Conference on Recommender Systems (RecSys '11).* (pp. 141–148) ACM: New York, doi=10.1145/2043932.2043960 http://doi.acm.org/10.1145/2043932.2043960.

Koda, T. (1996). Agents with faces: A study on the effects of personification of software agents. Master's Thesis, Massachusetts Institute of Technology, Boston, MA, USA.

Komiak, S. Y. X., & Benbasat, I. (2004). Understanding customer trust in agent-mediated electronic commerce, web-mediated electronic commerce and traditional commerce. *Information Technology and Management, 5*(1&2), 181–207.

Komiak, S. Y. X., & Benbasat, I. (2006). The effects of personalization and familiarity on trust and adoption of recommendation agents. *MIS Quarterly, 30*(4), 941–960.

Komiak, S. Y. X., Wang, W., & Benbasat, I. (2005). Trust Building in Virtual Salespersons Versus in Human Salespersons: Similarities and Differences. e-Service. *Journal, 3*(3), 49–63.

Koren, Y., Bell, R., & Volinsky, C. (2009). Matrix factorization techniques for recommender systems. *IEEE Computer, 42*(8), 30–37.

Krugman, H. E. (1962). On application of learning theory to TV copy testing. *Public Opinion Quarterly, 16*, 626–639.

Langlois, M. A., Petosa, R., & Hallam, J. S. (1999). Why do effective smoking prevention programs work? Student changes in social cognitive theory constructs. *Journal of School Health, 69*, 326–331.

Lautman, M. R., & Dean, K. J. (1983). Time compression of television advertising. In l. Percy & A. G. Woodside (Eds.), Advertising and consumer psychology (pp. 219–236). Lexington: Lexington Books.

Lascu, D.-N., Bearden, W. O., & Rose, R. L. (1995). Norm extremity and personal influence on consumer conformity. *Journal of Business Research, 32*(3), 201–213.

Latané, B. (1981). The psychology of social impact. *American Psychologist, 36*, 343–356.

Lazarsfeld, P., & Merton, R. K. (1954). Friendship as a social process: A substantive and methodological analysis. In M. Berger, T. Abel, & C. H. Page (Eds.), *Freedom and control in modern society* (pp. 18–66). New York: Van Nostrand.

Leventhal, L., Jones, S., & Trembly, G. (1966). Sex differences in attitude and behavior change under conditions of fear and specific instructions. *Journal of Experimental Social Psychology, 2*, 387–399.

Levine, R. V. (2003). Whom do we trust? Experts, honesty, and likability. In R. V. Levine (Ed.), *The power of persuasion* (pp. 29–63). Hoboken: Wiley.

Liang, T., Lai, H., & Ku, Y. (2006). Personalized content recommendation and user satisfaction: theoretical synthesis and empirical findings. *Journal of Management Information Systems, 23*(3), 45–70.

Lucente, M. (2000). Conversational interfaces for e-commerce applications. *Communications of the ACM, 43*(9), 59–61.

MacLachlan, J. (1982). Listener perception of time-compressed spokespersons. *Journal of Advertising Research, 22*(2), 47–51.

Maes, P., Guttman, R. H., & Moukas, A. G. (1999). Agents that buy and sell. *Communication of the ACM, 42*(3), 81–91.

Mahmood, T., Ricci, F., Venturini, A., & Höpken, W. (2008). Adaptive recommender systems for travel planning. In P. O'Connor, W. Höpken, & U. Gretzel (Eds.), *Information and communication technologies in tourism 2008* (pp. 1–11). Vienna: Springer.

Mahmood, T., Ricci, F., & Venturini, A. (2010). Improving recommendation effectiveness by adapting the dialogue strategy in online travel planning. *Information Technology and Tourism, 11*(4), 285–302.

Manning, C., Raghavan, P., & Schütze, H. (2008). *Introduction to information retrieval*. Cambridge: Cambridge University Press.

Mayer, R. C., Davis, J. H., & Schoorman, F. D. (1995). An integrative model of organizational trust. *Academy of Management Review, 20*, 709–734.

Mayer, R. E., Johnson, W. L., Shaw, E., & Sandhu, S. (2006). Constructing computer-based tutors that are socially sensitive: Politeness in educational software. *International Journal of Human-Computer Studies, 64*(1), 36–42.

McCroskey, J. C. (1970). The effects of evidence as an inhibitor of counter-persuasion. *Speech Monographs, 37*, 188–194.

McCroskey, J. C., & Mehrley, R. S. (1969). The effects of disorganization and nonfluency on attitude change and source credibility. *Speech Monographs, 36*, 13–21.

McGinty, L. & Smyth. B. (2002). Deep Dialogue vs Casual Conversation in Recommender Systems. In F. Ricci and B. Smyth (Eds.), *Proceedings of the workshop on personalization in eCommerce at the second international conference on adaptive hypermedia and web-based systems (AH 2002)*. (pp. 80–89). Universidad de Malaga, Malaga, Spain, Springer.

McGuire, W. J. (1961). Persistence of the resistance to persuasion induced by various types of prior defenses. *Journal of Abnormal and Social Psychology, 64*, 241–248.

McGuire, W. J. (1964). Inducing resistance to persuasion: Some contemporary approaches. In L. Berkowitz (Ed.), *Advances in experimental social psychology* (Vol. 1, pp. 191–229). New York: Academic.

McGuire, W. J. (1966). Attitudes and opinions. *Annual Review of Psychology, 17*, 475–514.

McGuire, W. J. (1968). The Nature of Attitudes and Attitude Change. In G. Lindzey & E. Aronson (Eds.), *Handbook of social psychology*. Reading: Addison-Wesley.

McGuire, W. J., & Papageorgis, D. (1961). The relative efficacy of various types of prior belief-defense in producing immunity against persuasion. *Journal of Abnormal and Social Psychology, 62*, 327–337.

McNee, S. M., Lam, S. K., Konstan, J. A., & Riedl, J. (2003). Interfaces for eliciting new user preferences in recommender systems. In user modeling 2003, LNCS 2702, 178–187. Springer.

Michener, H. A., DeLamater, J. D., & Myers, D. J. (2004). *Social psychology*. Wadsworth: Thomson Learning, Inc.

Mills, J., & Kimble, C. E. (1973). Opinion change as a function of perceived similarity of the communicator and subjectivity of the issue. *Bulletin of the Psychonomic Society, 2*, 35–36.

Mohr, L. A., & Bitner, M. J. (1995). The role of employee effort in satisfaction with service transactions. *Journal of Business Research, 32*(3), 239–252.

Moon, Y. (2002). Personalization and personality: some effects of customizing message style based on consumer personality. *Journal of Consumer Psychology, 12*(4), 313–326.

Moreno, R., Mayer, R. E., Spires, H. A., & Lester, J. C. (2001). The case for social agency in computer-based teaching: Do students learn more deeply when they interact with animated pedagogical agents? *Cognition and Instruction, 19*(2), 177–213.

Morkes, J., Kernal, H. K., & Nass, C. (1999). Effects of humor in task-oriented human-computer interaction and computer-mediated communication: A direct test of SRCT theory. *Human–Computer Interaction, 14*(4), 395–435.

Moulin, B., Irandoust, H., Belanger, M., & Desbordes, G. (2002). Explanation and argumentation capabilities: Towards the creation of more persuasive agents. *Artificial Intelligence Review, 17*, 169–222.

Moundridou, M., & Virvou, M. (2002). Evaluation the persona effect of an interface agent in a tutoring system. *Journal of Computer Assisted Learning, 18*(3), 253–261.

Munn, W. C., & Gruner, C. R. (1981). "Sick" Jokes, speaker sex, and informative speech. *Southern Speech Communication Journal, 46*, 411–418.

Murano, P. (2003). Anthropomorphic vs. non-anthropomorphic software interface feedback for online factual delivery. *Proceedings of the Seventh International Conference on Information Visualization*. Retrieved 1 Oct 2008 from http://portal.acm.org/citation.cfm?id=939634&dl=ACM&coll=portal.

Nanou, T., Lekakos, G., & Fouskas, K. (2010). The effects of recommenders' presentation on persuasion and satisfaction in a movie recommender system. *Multimedia Systems, 16*, 219–230.

Nass, C., & Brave, S. (2005). *Wired for speech: How voice activates and advances the human-computer relationship*. Cambridge: MIT Press.

Nass, C., Fogg, B. J., & Moon, Y. (1996). Can computers be teammates? *International Journal of Human-Computer Studies, 45*(6), 669–678.

Nass, C., Isbister, K., & Lee, E. –J. (2000). Truth is beauty: Researching embodied conversational agents. In J. Cassell, J. Sullivan, S. Prevost, & E. Churchill (Eds.), *Embodied conversational agents* (pp. 374–402). Cambridge, MA: MIT Press.

Nass, C., & Moon, Y. (2000). Machines and mindlessness: Social responses to computers. *Journal of Social Issues, 56*(1), 81–103.

Nass, C., Moon, Y., & Green, N. (1997). Are computers gender-neutral? Gender stereotypic responses to computers. *Journal of Applied Social Psychology, 27*(10), 864–876.

Nguyen, H., Masthoff, J. & Edwards. P. (2007). Persuasive effects of embodied conversational agent teams. *Proceedings of 12th International Conference on Human-computer Interaction* (pp. 176–185). Springer-Verlag: Beijing, China Berlin.

Nowak, K. (2004). The influence of anthropomorphism and agency on social judgment in virtual environments. *Journal of Computer-Mediated Communication 9* (2).

Nowak, K. L., & Biocca, F. (2003). The effect of the agency and anthropomorphism on user's sense of telepre4sence, copresence, and social presence in virtual environments. *Presence: Teleperators and Virtual Environments, 12*(5), 481–494.

Nowak, K., & Rauh, C. (2005). The influence of the avatar on online perceptions of anthropomorphism, androgyny, credibility, homophily, and attraction. *Journal of Computer-Mediated Communication, 11* (1).

Ochi, P., Rao, S., Takayama, L., & Nass, C. (2010). Predictors of user perceptions of web recommender systems: How the basis for generating experience and search product recommendations affects user responses. *International Journal of Human-Computer Studies, 68*, 472–482.

O'Hare, M. A., Phillips, J. G., & Moss, S. (2009). The effect of contextual and personal factors on the use of probabilistic recommenders in E-markets. *The Ergonomics Open Journal, 2*, 207–216.

O'Keefe, D. J. (1997). Standpoint explicitness and persuasive effect: A meta-analytic review of the effects of varying conclusion articulation in persuasive messages. *Argumentation and Advocacy, 34*, 1–12.

O'Keefe, D. J. (1998). Justification explicitness and persuasive effect: A meta-analytic review of the effects of varying support articulation in persuasive messages. *Argumentation and Advocacy, 35*, 61–75.

O'Keefe, D. J. (2002). *Persuasion: Theory and research*. Thousand Oaks: Sage Publications.

Olshavsky, R. W. (1985). Toward a more comprehensive theory of choice. *Advances in Consumer Research, 12*, 465–470.

Omoto, A. M., Snyder, M., & Martino, S. C. (2000). Volunteerism and the life course: Investigating age-related agendas for action. *Basic and Applied social Psychology, 22*, 181–197.

Ozok, A. A., Fan, Q., & Norcio, A. F. (2010). Design guidelines for effective recommender system interfaces based on a usability criteria conceptual model: Results from a college student population. *Behaviour and Information Technology, 29*(1), 57–83.

Papageorgis, D. (1968). Warning and persuasion. *Psychological Bulletin, 70*, 271–282.

Patzer, G. L. (1983). Source credibility as a function of communicator physical attractiveness. *Journal of Business Research, 11*(2), 229–241.

Pechmann, C. (1992). Predicting when two-sided Ads will be more effective than one-sided Ads. *Journal of Marketing Research, 24*, 441–453.

Pereira, R. E. (2000). Optimizing human-computer interaction for the electronic commerce environment. *Journal of Electronic Commerce Research, 1*(1), 23–44.

Perloff, R. M. (2003). *The dynamics of persuasion* (2nd ed.). Mahwah: Lawrence Erlbaum Associates.

Petty, R. E., & Cacioppo, J. T. (1977). Forewarning, cognitive responding, and resistance to persuasion. *Journal of Personality and Social Psychology, 35*, 645–655.

Petty, R. E., & Cacioppo, J. T. (1981). *Attitudes and persuasion: Classic and contemporary approaches*. Dubuque: William C. Brown.

Petty, R. E., & Cacioppo, J. T. (1986). *Communication and persuasion: Central and peripheral routes to attitude change*. New York: Springer-Verlag.

Petty, R. E., & Cacioppo, J. T. (1990). Involvement and persuasion: Tradition versus integration. *Psychological Bulletin, 107*, 367–374.

Pfau, M., Holbert, R. L., Zubric, S. J., Pasha, N. H., & Lin, W.-K. (2000). Role and influence of communication modality in the process of resistance to persuasion. *Media Psychology, 2*, 1–33.

Pittam, J. (1994). *Voice in social interaction: An interdisciplinary approach*. Thousand Oaks: Sage.

Pu, P., & Chen, L. (2007). Trust-inspiring explanation interfaces for recommender systems. *Knowledge-Based Systems, 20*, 542–556.

Pu, P., Chen, L. & Hu, R. (2011). A user-centric evaluation framework for recommender systems. In *Proceedings of the 2011 ACM Conference on Recommender Systems (RecSys'11)*. pp. 157–164.

Qiu, L. (2006). Designing social interaction with animated avatars and speech output for product recommendation agents in electronic commerce. Doctoral Thesis, University of British Columbia, Vancouver.

Reeves, B., & Nass, C. (1996). *The media equation: How people treat computers, television, and new media like real people and places*. New York: CSLI.

Rhoads, K. V., & Cialdini, R. B. (2002). The business of influence. In J. P. Dillard & M. Pfau (Eds.), *Persuasion handbook: Developments in theory and practice* (pp. 513–542). London, United Kingdom: Sage.

Rhodes, N., & Wood, W. (1992). Self-esteem and intelligence affect influence ability: The mediating role of message reception. *Psychological Bulletin, 111*, 156–171.

Ricci, F. (2011). Mobile recommender systems. *Information Technology and Tourism, 12*(3), 205–231.

Ricci, F., & Nguyen, Q. N. (2007). Acquiring and revising preferences in critique-based mobile recommender system. *IEEE Intelligent Systems, 22*(3), 22–29. doi:10.1109/MIS.2007.43.

Sampson, E. E., & Insko, C. A. (1964). Cognitive consistency and performance in the autokinetic situation. *Journal of Abnormal and Social Psychology, 68*, 184–192.

Sawyer, A. (1973). The effects of repetition of refutational and supportive advertising appeals. *Journal of Marketing Research, 10*, 23–33.

Schafer, J. B. (2005). DynamicLens: A dynamic user-interface for a meta-recommendation system. Workshop: Beyond Personalization 2005, IUI'05, San Diego, CA.

Schafer, J. B., Knostan, J. A., & Riedl, J. (2002). Meta-recommendation systems: user-controlled integration of diverse recommendations. Paper Presented at the 11th International Conference on Information and Knowledge Management, McLean, VA, Novemeber 2002.

Schafer, J.B., Konstan, J.A., & Reidl, J. (2004). The view through MetaLens: Usage patterns for a meta recommendation system. *IEEE Software*.

Schliesser, H. F. (1968). Informatio transmission and ethos of a speaker using normal and defective speech. *Central States Speech Journal, 19*, 169–174.

Sebastian, R. J., & Bristow, D. (2008). Formal or informal? The impact of style of dress and forms of address on business students' perceptions of professors. *Journal of Education for Business, 83*(4), 196–201.

Self, C. S. (1996). Credibility. In M. B. Salwen & D. W. Stacks (Eds.), *An integrated approach to communication theory and research* (pp. 421–441). Mahwah: Lawrence Erlbaum.

Sénécal, S., & Nantel, J. (2003). Online influence of relevant others: A framework. *Proceedings of the Sixth International Conference on Electronic Commerce Research (ICECR-6)*, Dallas, Texas.

Sénécal, S., & Nantel, J. (2004). The influence of online product recommendations on consumers' online choices. *Journal of Retailing, 80*(2), 159–169.

Shavitt, S., & Brock, T. C. (1994). *Persuasion: Psychological Insights and Perspectives*. Needham Heights: Allyn and Bacon.

Shimazu, H. (2002). Expertclerk: A conversational case-based reasoning tool for developing salesclerk agents in e-commerce webshops. *Artificial Intelligence Review, 18*(3–4), 223–244.

Simon, H. (1959). Theories of decision making in economics and behavioural science. *American Economic Review, 49*(3), 253–283.

Sinha, R. & Swearingen, K. (2001). Comparing recommendations made by online systems and friends. *Proceedings of the 2nd DELOS Network of Excellence Workshop on Personalization and Recommender Systems in Digital Libraries*, Dublin Ireland, June 18–20.

Sinha, R. & Swearingen, K. (2002). The role of transparency in recommender systems. In CHI 2002 extended abstracts on human factors in computing systems, Minneapolis, MN, April 830–831.

Smith, R. E., & Hunt, S. D. (1978). Attributional processers and effects in promotional situations. *Journal of Consumer Research, 5*, 149–158.

Smith, D., Menon, S., & Sivakumar, K. (2005). Online peer and editorial recommendations, trust, and choice in virtual markets. *Journal of Interactive Marketing, 19*(3), 15–37.

Snyder, M., & Rothbart, M. (1971). Communicator attractiveness and opinion change. *Canadian Journal of Behavioural Science, 3*, 377–387.

Sparks, J. R., Areni, C. S., & Cox, K. C. (1998). An investigation of the effects of language style and communication modality on persuasion. *Communication Monographs, 65*(2), 108–125.

Spiekermann, S. (2001). Online information search with electronic agents: drivers, impediments, and privacy issues, (Unpublished doctoral dissertation), Humboldt University Berlin, Berlin, Germany.

Sproull, L., Subramani, M., Kiesler, S., Walker, J. H., & Waters, K. (1996). When the interface is a face. *Human-Computer Interaction, 11*(1), 97–124.

Struckman-Johnson, D., & Struckman-Johnson, C. (1996). Can you say condom? It makes a difference in fear-arousing AIDS prevention public service announcements. *Journal of Applied Social Psychology, 26*, 1068–1083.

Suh, K. S. (1999). Impact of communication medium on task performance and satisfaction: an examination of media-richness theory. *Information and Management, 35*, 295–312.

Sutcliffe, A. G., Ennis, M., & Hu, J. (2000). Evaluating the effectiveness of visual user interfaces for information retrieval. *International Journal of Human–Computer Studies, 53*, 741–763.

Swartz, T. A. (1984). Relationship between source expertise and source similarity in an advertising context. *Journal of Advertising, 13*(2), 49–55.

Swearingen, K., & Sinha, R. (2001). Beyond algorithms: An HCI perspective n recommender systems. *Proceedings of the ACM SIGIR 2001 Workshop on Recommender Systems*, New Orleans: Louisiana.

Swearingen, K. & Sinha, R. (2002). Interaction design for recommender systems. Designing Interactive Systems 2002. ACM, 2002.

Tamborini, R., & Zillmann, D. (1981). College students' perceptions of lecturers using humor. *Perceptual and Motor Skills, 52*, 427–432.

Taylor, P. M. (1974). An experimental study of humor and ethos. *Southern Speech Communication Journal, 39*, 359–366.

Teppan, E. C., & Felfernig, A. (2009). Impacts of decoy elements on result set evaluations in knowledge-based recommendation. *International Journal of Advanced Intelligence Paradigms, 1*, 358–373.

Teppan, E. C. & Felfernig, A. (2010). Minimization of product utility estimation errors in recommender result set evaluations. *Proceedings of the International Conference on Web Intelligence and Intelligent Agent Technology* (pp. 20–27) IEEE.

Teppan, E.C. & Zanker, M. (2012). Decision Biases in Recommender Systems, Technical report, Institut für Angewandte Informatik, Universität Klagenfurt.

Tintarev, N., & Masthoff, J. (2007). Effective explanations of recommendations: User-centered design. *Proceedings of the ACM Conference on Recommender Systems* (pp. 153–156) Minneapolis, US.

Tintarev, N. & Masthoff, J. (2011). Designing and evaluating explanations for recommender systems, In F. Ricci et al. (Eds.), *Recommender Systems Handbook* (pp. 479–510) Springer.

Trevino, L. K., Daft, R. L., & Lengel, R. H. (1990). Understanding managers' media choices: A symbolic interactionist perspective. In J. Fulk & C. Steinfield (Eds.), *Organizations and communication technology* (pp. 71–94). Newbury Park: Sage.

Tversky, A., & Kahnemann, D. (1986). Rational choice and the framing of decisions. *Journal of Business, 59*, 251–278.

Tzeng, J.-Y. (2004). Toward a more civilized design: Studying the effects of computers that apologize. *International Journal of Human-Computer Studies, 61*(3), 319–345.

Urban, G., Sultan, F., & Qualls, W. (1999). Design and evaluation of a trust based advisor on the Internet. Available at http://ebiz.mit.edu/research/papers/123%20Urban,%20Trust%20Based%20Advisor.pdf.

Van Slyke, C., & Collins, R. (1996). Trust between users and their intelligent agents. Annual Meeting of the Decision Sciences Institute. Orlando, FL, Nov. 24–26.

Vig, J., Sen, S. & Riedl, J. (2009). Tagsplanations: Explaining recommendations using tags. *Proceedings 14th International Conference on Intelligent User Interfaces*, pp. 47–56.

Wang, W. (2005). Design of trustworthy online recommendation agents: Explanation facilities and decision strategy support. Doctoral Thesis, University of British Columbia, Vancouver.

Wang, W., & Benbasat, I. (2005). Trust in and adoption of online recommendation agents. *Journal of the Association for Information Systems, 6*(3), 72–101.

Wang, W., & Benbasat, I. (2007). Recommendation agents for electronic commerce: effects of explanation facilities on trusting beliefs. *Journal of Management Information Systems, 23*(4), 217–246.

Wang, W., & Benbasat, I. (2008). Analysis of trust formation in online recommendation agents. *Journal of Management Information Systems, 24*(4), 249–273.

Wang, Y. D., & Emurian, H. H. (2005). An overview of online trust: Concepts, elements and implications. *Computers in Human Behavior, 21*(1), 105–125.

Weizenbaum, J. (1966). Computational linguistics. *Communications of the ACM, 9*(1), 36–45.

West, P. M., Ariely, D., Bellman, S., Bradlow, E., Huber, J., Johnson, E., et al. (1999). Agents to the rescue? *Marketing Letters, 10*(3), 285–300.

Wilson, E. J., & Sherrell, D. L. (1993). Source effects in communication and persuasion research: A meta-analysis of effect size. *Journal of the Academy of Marketing Science, 21*, 101–112.

Wolf, S., & Bugaj, A. M. (1990). The social impact of courtroom witnesses. *Social Behaviour, 5*(1), 1–13.

Xiao, B., & Benbasat, I. (2007). E-Commerce product recommendation agents: Use, characteristics, and impact. *MIS Quarterly, 31*(1), 137–209.

Yoo, K. -H. (2010). Creating more credible and likable recommender systems. (Unpublished doctoral dissertation). Texas A&M University, College Station, USA.

Yoo, K.-H., Lee, K. S., & Gretzel, U. (2007). The role of source characteristics in eWOM: What makes online travel reviewers credible and likeable? In M. Sigala, L. Mich, J. Murphy, and A. Frew (Eds.), *Information and communication technologies in tourism* (pp. 23–34) Springer.

Yoo, K.-H., & Gretzel, U. (2008). The influence of perceived credibility on preferences for recommender systems as sources of advice. *Information Technology and Tourism, 10*(2), 133–146.

Yoo, K. -H, & Gretzel, U. (2009). The influence of virtual representatives on recommender system evaluation. *Proceedings of the 15th Americas Conference on Information Systems*, San Francisco, California.

Yoon, S. N., & Lee, Z. (2004). The impact of the Web-based product recommendation systems from previous buyers on consumers' purchasing behavior. Paper presented at the tenth Americas Conference on Information Systems, New York, New York, August, 2004.

Zanker, M., Bricman, M., Gordea, S., Jannach, D. & Jessenitschnig, M. (2006). Persuasive online-selling in quality & taste domains. E-Commerce and Web Technologies. Lecture Notes in Computer Science, 4082/2006, 51–60, doi: 10.1007/11823865_6.

Zanker M., Bricman M. & Jessenitschnig M. (2009). Cost-efficient development of virtual sales assistants. *Proceedings of the 2nd International Symposium on Intelligent Interactive Multimedia Systems and Services (KES IIMSS)*, Mogliano Veneto, Italy, 2009, 1–11.

Zanker M., Fuchs M., Höpken W., Tuta M. & Müller N. (2008). Evaluating recommender systems in tourism—A case study from Austria. In P. O'Connor et al. (Eds.), Information and communication technologies in tourism, Springer, 2008, pp. 24–34.

Zanker, M., & Jessenitschnig, M. (2009). Case-studies on exploiting explicit customer requirements in recommender systems. *User Modeling and User-Adapted Interaction, Springer, 19*(1–2), 133–166. doi:10.1007/s11257-008-9048-y.

Zanker, M., Jessenitschnig, M., & Schmid, W. (2010). Preference reasoning with soft constraints in constraint-based recommender systems. *Constraints, Springer, 15*(4), 574–595. doi: 10.1007/s10601-010-9098-8.

Zanker, M. & Ninaus, D. (2010). Knowledgable explanations for recommender systems. *Proceedings of the IEEE/WIC/ACM international conference on web intelligence and intelligent agent technology (WI/IAT), IEEE* (pp. 657–660). doi:10.1109/WI-IAT.2010.131.

Zuckerman, M. (1979). *Sensation seeking: Beyond the optimal level of arousal.* Lawrence Erlbaum: Hillsdale.